Engineering Analysis with COSMOSWorks

SolidWorks 2004 / COSMOSWorks 2004

Paul M. Kurowski Ph.D., P.Eng.

ISBN: 1-58503-181-X

Design Generator, Inc.

Schroff Development Corporation

www.schroff.com
www.schroff-europe.com

Trademarks and Disclaimer

SolidWorks and its family of products are registered trademarks of Dassault Systemes. COSMOSWorks and COSMOSDesignSTAR are registered trademarks of Structural Research & Analysis Corporation. Microsoft Windows and its family products are registered trademarks of the Microsoft Corporation.

Every effort has been made to provide an accurate text. The author and the manufacturers shall not be held liable for any parts developed with this book or held responsible for any inaccuracies or errors that appear in the book.

Copyright © 2004 by Paul M. Kurowski

All rights reserved. This document may not be copied, photocopied, reproduced, transmitted, or translated in any form or for any purpose without the express written consent of the publisher, Schroff Development Corporation.

Acknowledgements

Writing this book, the first of its kind for COSMOSWorks software, was a substantial effort that would not have been possible without the help and support of my professional colleagues and friends. I would like to acknowledge the following:

- SolidWorks Corporation — Suchit Jain, Sanjay Deo
- Javelin Technologies, Inc. — John Carlan, Ted Lee, Bill McEachern, Joseph Vera, Karen Zapata
- International Internet Consultants — Lisa Syed

Also, I would like to thank the students attending my various training courses in Finite Element Analysis for their questions and comments that helped to shape the unique approach this book is taking. I thank my wife Elzbieta for her support and encouragement that made this book possible.

About the Author

Dr. Paul Kurowski obtained his M.Sc. and Ph.D. in Applied Mechanics from Warsaw Technical University, Warsaw, Poland. He completed postdoctoral work at Warsaw Technical University, Kyoto University, and the University of Western Ontario. Paul is the President of Design Generator Inc., a consulting firm with expertise in Product Development and training in Computer Aided Engineering methods. His teaching experience includes Finite Element Analysis (FEA), Machine Design, Mechanics of Materials, and Solid Modeling for various colleges, universities, professional organizations, and industries. He has published many technical papers and created and taught professional development seminars in the field of Finite Element Analysis for the Society of Automotive Engineers, the Association of Professional Engineers of Ontario, the Parametric Technology Corporation (PTC), Rand Worldwide, SolidWorks Corporation, and Javelin Technologies.

Paul is a member of the Association of Professional Engineers of Ontario and the Society of Automotive Engineers. His professional interests revolve around finding the best ways of using FEA as a design tool for faster and more effective product development processes where computer models replace physical prototypes. FEA competes for his time dedicated to his family and to his passion for Latin America and mountaineering. Paul Kurowski can be contacted at: pkurowski@rogers.com.

About the cover

The image on the cover presents von Mises stress results of a hollow bracket with unsuppressed small rounds from the hollow bracket exercise. See chapter 15 for details.

Table of contents

Before You Start — 1
   Notes on hands-on exercises — 1
   Prerequisites — 1
   Windows 2000/XP terminology — 2

1: Introduction — 3
   What exactly is Finite Element Analysis? — 3
   Who should use FEA? — 3
   FEA: another design tool — 4
   FEA: based on CAD models — 4
   FEA: concurrent with the design process — 4
   Limitations of "FEA for design engineers" — 4
   Objective of "FEA for design engineers" — 5
   What is COSMOSWorks? — 6
   Integration between COSMOSWorks and SolidWorks — 6
   What are the fundamental steps in an FEA project? — 6
   Errors in FEA — 10
   A closer look at finite elements — 10
   Degrees of freedom — 16
   What is really calculated in FEA? — 16
   How do we interpret FEA results? — 16
   Von Mises stress — 17
   Principal stresses: P1, P2, and P3 — 19
   Units of measurements — 20
   Using on-line Help — 21
   Limitations of COSMOSWorks — 22

2: Analysis of a rectangular plate with a hole — 27
   Objectives — 27
   Project description — 27
   Procedure — 28

3: Analysis of an L-bracket — 65
   Objectives — 65

Project description	*65*
Procedure	*66*

4: Analysis of a support bracket — 81

Objectives	*81*
Project description	*81*
Procedure	*81*

5: Analysis of a Link — 97

Objectives	*97*
Project description	*97*
Procedure	*98*

6: Analysis of a tuning fork — 105

Objectives	*105*
Project description	*105*
Procedure	*106*

7: Analysis of crossing pipes — 113

Objectives	*113*
Project description	*113*
Procedure	*114*

8: Analysis of a radiator assembly — 119

Objectives	*119*
Project description	*119*
Procedure	*120*

9: Analysis of a hanger assembly — 133

Objectives	*133*
Project description	*133*
Procedure	*137*

10: Analysis of two cylinders in contact — 145

Objectives	*145*
Project description	*145*
Procedure	*146*

11: Analysis of a bi-metal beam — 151

Objectives	*151*
Project description	*151*
Procedure	*152*

12: Analysis of a pipe with cooling fins   **157**
Objective   *157*
Project description   *157*
Procedure   *157*

13: Analysis of an L-beam   **163**
Objectives   *163*
Project description   *163*
Procedure   *164*

14: Optimization of a plate in bending   **169**
Objectives   *169*
Project description   *169*
Procedure   *169*

15: Analysis of a hollow bracket   **181**
Objectives   *181*
Project description   *181*
Procedure   *182*
Why are upgradeable elements called p-elements?   *184*
Why is the p-convergence process called a p-Adaptive solution, and what exactly does "Adaptive" mean?   *184*
Procedure, continued   *184*

16: Analysis of a tapered block   **191**
Objective   *191*
Project description   *191*
Procedure   *192*

17: Miscellaneous topics   **201**
Selecting the automesher   *201*
Mesh quality and mesh degeneration   *201*
Solvers and solvers options   *202*
Displaying mesh in a results plot   *205*
Creating automatic reports   *206*
Using e-drawings for result presentation   *208*
Defining non-uniform loads   *208*
Defining bearing loads   *209*

18: Selected advanced topics — 211
Frequency analysis with pre-load — *211*
Large deformations contact — *214*
Shrink fit and inertial relief — *217*

19: Implementation of FEA into the design process — 221
Positioning CAD and FEA activities — *221*
Major steps in an FEA project — *222*
Progress checkpoints in an FEA project — *226*
Structure of an FEA report — *227*

20: Glossary — 229

21: FEA Resources — 237

Before You Start

Notes on hands-on exercises

This book goes beyond a standard software manual because its unique approach concurrently introduces you to COSMOSWorks software and the fundamentals of Finite Element Analysis through hands-on exercises. We recommend that you study the exercises in the order presented in the text. As you go through the exercises, you will notice that explanations and steps described in detail in earlier exercises are not repeated later. Each subsequent exercise assumes familiarity with FEA fundamentals and software functions discussed in previous exercises. Each exercise builds on the skills, experience, and understanding of the problem gained from the previous exercises. An exception to the above is chapter 19, *Implementation of FEA into the design process*, which stands independent of the hands-on exercises.

All exercises use SolidWorks models of parts or assemblies, which you can download from http://www.schroff1.com/. For your reference, we also provide exercises in ready-to-run form. Those, however, should be treated as a "last resort" as we encourage you to complete exercises without this help.

This book is not intended to replace regular software manuals. While you are guided through the specific exercises, not all of the software functions are explained. We encourage you to explore each exercise beyond its description. Investigate other options, other menu choices, and ways to present results. You will soon discover that the same simple logic applies to all functions in COSMOSWorks software.

All SolidWorks file names appear in CAPITAL letters, even though the actual file name may use a combination of small and capital letters. Selected menu items and COSMOSWorks commands appear in **bold**. Folder names and icons appear in *italics*.

Prerequisites

We assume that you have the following prerequisites:

- Understanding of the concepts of the Mechanics of Materials
- Experience with parametric, solid modeling using SolidWorks software
- Familiarity with the Windows Operating System

Windows 2000/XP terminology

The mouse pointer plays a very important role in executing various commands and providing user feedback. Use the mouse pointer to execute commands, select geometry, and invoke pop-up menus. We will use Windows terminology when referring to mouse-pointer actions.

Item	Description
Click	Press and release the left mouse button.
Double-click	Double press and release the left mouse button.
Click-inside	Press the left mouse button. Wait a second, and then press the left mouse button inside the pop-up menu or text box. Use this technique to modify the names of folders and icons in COSMOSWorks Manager.
Drag	Use the mouse to point to an object. Press and hold the left mouse button down. Move the mouse pointer to a new location. Release the left mouse button.
Right-click	Press and release the right mouse button. A pop-up menu is displayed. Use the left mouse button to select a menu command.
Tool/Tip	Position the mouse pointer over an icon. The command is displayed below the mouse pointer.
Mouse pointer feedback	Position the mouse pointer over various areas of the model. The cursor provides the same feedback as in SolidWorks.

1: Introduction

What exactly is Finite Element Analysis?

Finite Element Analysis, commonly called FEA, is a method of numerical analysis. FEA has been found useful for solving problems in many engineering disciplines, such as machine design, acoustics, electromagnetism, soil mechanics, fluid dynamics, and many others. In mathematical terms, FEA is a numerical technique used for solving field problems described by a set of partial differential equations.

In mechanical engineering, FEA is widely used for solving structural, vibration, and thermal problems. However, FEA is not the only available tool for numerical analysis. Other numerical methods include the Finite Difference Method, the Boundary Element Method, or the Finite Volumes Method to mention just a few. However, due to its versatility and high numerical efficiency, FEA has come to dominate the engineering analysis software market while other methods have been relegated to niche applications. You can use FEA to analyze any shape; FEA works with different levels of geometry idealization and provides results with the desired accuracy. When implemented into modern commercial software, FEA theory, numerical problem formulation, and solution methods become completely transparent to users.

Who should use FEA?

As a powerful tool for engineering analysis, FEA is used to solve problems ranging from very simple to very complex. Design engineers use FEA during the product development process to analyze the design-in-progress. Time constraints and limited availability of product data call for many simplifications of the analysis models. At the other end of scale, specialized analysts implement FEA to solve very advanced problems, such as vehicle crash dynamics, hydro forming, or air bag deployment.

This book focuses on how design engineers use FEA as a design tool. Therefore, we first need to explain what exactly distinguishes FEA as performed by design engineers from "regular" FEA. We will then highlight the most essential FEA characteristics for design engineers as opposed those for analysts.

FEA: another design tool

For design engineers, FEA is one of many design tools among CAD, spreadsheets, catalogs, data bases, hand calculations, text books, etc. that are all used in the design process.

FEA: based on CAD models

Since design is nowadays conducted using CAD tools, a CAD model is the starting point to analysis. Since CAD models are used for describing geometric information for FEA, the very important differences between CAD geometry and FEA models must always be appreciated. We will discuss them in later chapters.

FEA: concurrent with the design process

Since FEA is a design tool, it should be used concurrently with the design process. It should keep up with, or better yet, *drive* the design process. Analysis iterations must be performed fast and, since results are used to make design decisions, the results must be reliable, even though only limited input data may be available for analysis conducted early in the design process.

Limitations of "FEA for design engineers"

As you can see, FEA used in the design environment must meet high requirements. It must be fast and accurate; even though it is in the hands of design engineers and not FEA specialists. An obvious question arises: would it be better to have a dedicated specialist perform FEA and let design engineers do what they do best—designing new products? The answer depends on the size of organization, type of products, company organization and culture, and many other tangible and non-tangible factors. A general consensus is that design engineers should handle relatively simple types of analysis, but do that quickly and reliably. More complex types of analyses, too complex and too time consuming to be executed concurrently with the design process, are usually better handled either by a dedicated analyst or contracted out to specialized consultants.

Objective of "FEA for design engineers"

The ultimate objective of using the FEA as a design tool is to change the design process from several iterative cycles of "design, prototype, test" into a streamlined process where prototypes are not used as design tools. The use of prototypes is limited to final design verification. With the use of FEA, design iterations are moved from the physical space of prototyping and testing into the virtual space of computer simulations (figure 1-1).

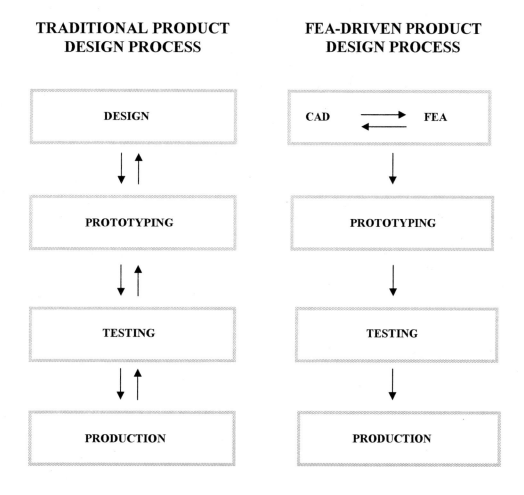

Figure 1-1: Traditional vs. FEA-driven product development

Traditional product development needs prototypes to support design in progress. The process in FEA-driven product development uses numerical models, rather than physical prototypes, to drive development. In an FEA-driven product, the design process prototype is no longer a part of the iterative design loop.

What is COSMOSWorks?

COSMOSWorks is a commercial implementation of FEA, capable of solving problems commonly found in design engineering, such as analysis of deformations, stresses, natural frequencies, heat flow, etc. COSMOSWorks addresses the needs of design engineers. It belongs to the family of engineering analysis software products developed by the Structural Research & Analysis Corporation (SRAC). SRAC was established in 1982 and since its inception has contributed innovations that have had a significant impact on the evolution of FEA. In 1995, SRAC entered the emerging mainstream FEA software market, partnering with SolidWorks Corporation and creating COSMOSWorks, one of the first SolidWorks Gold Products, which became the top-selling analysis solution for SolidWorks Corporation. The commercial success of COSMOSWorks integrated with SolidWorks CAD software resulted in the acquisition of SRAC in 2001 by Dassault Systemes, parent of SolidWorks Corporation. In 2003, SRAC operations merged with SolidWorks.

Integration between COSMOSWorks and SolidWorks

COSMOSWorks is tightly integrated with SolidWorks CAD software and uses SolidWorks for creating and editing model geometry. SolidWorks is a parametric, solid, feature-driven CAD system. As opposed to many other CAD systems that were originally developed in a UNIX environment and only later ported to Windows, SolidWorks CAD has been developed from the very beginning specifically for the Windows Operating System.

In summary, although the history of the family of COSMOS FEA products dates back to 1982, COSMOSWorks has been specifically developed for Windows and takes full advantage this of deep integration between SolidWorks and Windows representing the state-of-the-art in the engineering analysis software.

What are the fundamental steps in an FEA project?

The starting point for any COSMOSWorks project is a SolidWorks model, a part or an assembly, representing the object that needs to be analyzed. To this model, we assign material properties, and define loads, supports, and restraints. Next, as is always the case using any FEA-based analysis tool, we need to split the geometry into relatively small and simply shaped entities, called finite elements. The elements are called "finite" to emphasize the fact that they are not infinitesimally small, but only reasonably small in comparison to the overall model size. Creating finite elements is commonly called meshing. When working with finite elements, the COSMOSWorks solver approximates the wanted solution (for example, deformations or stresses) for the entire model with the assembly of simple solutions for individual elements.

From the perspective of FEA software, each application of FEA requires three steps:

- Preprocessing of the FEA model, where type of analysis, material properties, loads and restraints are defined and the model is split into finite elements
- Solution for computing wanted results
- Post-processing for results analysis

We will follow the above three steps every time we use COSMOSWorks.

From the perspective of FEA methodology, we can list the following FEA steps:

- Building the mathematical model
- Building the finite element model
- Solving the finite element model
- Analysis of results

The following subsections discuss these steps.

Building the mathematical model

The starting point to analysis with COSMOSWorks is a SolidWorks model. This geometry needs to be meshable into a correct and reasonably small finite element mesh. This requirement of mesh-ability has very important implications. We need to ensure that the CAD geometry will indeed mesh and that the produced mesh will provide the correct solution of the data of interest, such as displacements, stresses, temperature distribution, etc. This necessity often requires modifications to the CAD geometry, which can take the form of defeaturing, idealization and/or clean-up, described below:

Term	Description
Defeaturing	The process of removing geometry features deemed insignificant for analysis, such as external fillets, logos, etc.
Idealization	A more aggressive exercise that may depart from solid CAD geometry, for example, representing thin walls with surfaces
Clean-up	Sometimes required because the meshable geometry must satisfy much higher quality requirements than those required Solid Modeling. For cleanup, we can use CAD quality-control tools to check for problems, like sliver faces, multiple entities, etc. that could be tolerated in the CAD model, but would make subsequent meshing difficult or impossible

It is important to mention that we do not always simplify the CAD model with the sole objective of making it meshable. Often, we simplify a model that would mesh correctly "as is", but the resulting mesh would be too large and, consequently, the analysis would run too slowly. Geometry modifications allow for a simpler mesh and shorter computing times. Also, geometry preparation is not always required at all. Successful meshing depends as much on the quality of geometry submitted for meshing as on the sophistication of the meshing tools implemented in the FEA software.

Having prepared a meshable, but not yet meshed geometry, we now define material properties, loads, supports and restraints, and provide information on the type of analysis that we wish to perform. This procedure completes the creation of a mathematical model (figure 1-2). Notice that the process of creating the mathematical model is not FEA-specific. FEA has not yet entered the picture.

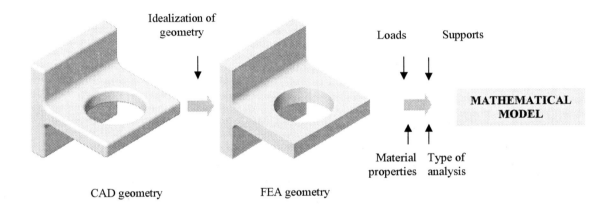

Figure 1-2: Building the mathematical model

The process of creating a mathematical model consists of the idealization of CAD geometry (here removing external rounds), definition of loads, supports, material properties, and the type of analysis (e.g., static, thermal, or modal) that we wish to perform.

Building the finite element model

The mathematical model now needs to be split into finite elements, through a process of discretization, more commonly known as meshing. Discretization visually manifests itself as the meshing of geometry. However, loads and supports are also discretized and, once the model has been meshed, the discretized loads and supports are applied to nodes of the finite element mesh.

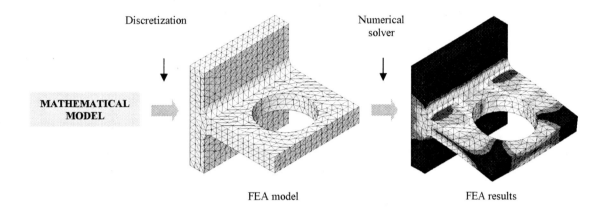

Figure 1-3: Building the finite element model

The mathematical model is discretized into a finite element model. This completes the pre-processing phase. The FEA model is then solved with one of the numerical solvers available in COSMOSWorks. The analysis of results, called post-processing, concludes the FEA.

Solving the finite element model

Having created the finite element model, we now use a solver provided in COSMOSWorks to produce the desired data of interest (figure 1-3).

Analyzing the results

Often the most difficult step of all, the analysis of the results provides very detailed results data. This information can be presented in almost any format. Proper interpretation of results requires that we appreciate all simplifications and errors introduced in the first three steps: building the mathematical model, building the finite element model, and solving the finite element model.

Errors in FEA

The process illustrated in figures 1-2 and 1-3 introduces unavoidable errors. Formulation of a mathematical model introduces modeling errors, also called idealization errors. Discretization of the mathematical model introduces discretization errors and solution introduces numerical errors. Of these three types of errors, only discretization errors are specific to FEA. Therefore, only discretization errors can be controlled using FEA methods. Modeling errors, affecting mathematical model, are introduced before FEA is utilized and can only be controlled by using correct modeling techniques. Solution errors, the accumulation of round-off errors in the solver, are difficult to control, but fortunately are usually low.

A closer look at finite elements

The discretization process, better know as meshing, splits continuous mathematical models into finite elements. The type of element created in this process depends on the type of geometry meshed, the type of analysis that needs to be executed, as well as our own preferences. COSMOSWorks offers two types of elements: tetrahedral solid elements, for meshing solid geometry, and shell elements, for meshing surface geometry.

Before proceeding we need to clarify an important terminology issue. What in CAD terminology we call solid geometry, in FEA is called volumes. Solid elements are then used to mesh those volumes. Note that the term *solid* has different meanings when it is used in "*solid* geometry" and in "*solid* element".

Solid elements

The type of geometry that is most often used for analysis with COSMOSWorks is solid CAD geometry. Meshing of this geometry is accomplished with tetrahedral solid elements commonly, called *tets* in FEA jargon. The tetrahedral solid elements in COSMOSWorks can either be first order elements (draft quality) or second order elements (high quality). The user decides whether to use draft quality or high quality elements for meshing. However, as we will soon prove, only high quality elements should be used for an analysis of any importance. The difference between first and second order tetrahedral elements is illustrated in figure 1-4.

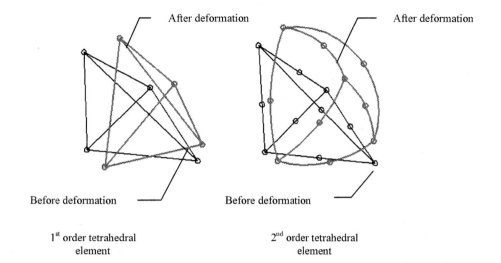

Figure 1-4: Differences between first and second order tetrahedral elements

First and the second order tetrahedral elements are shown before and after deformation. Note that the deformed faces of the second order element may assume either convex or concave shape.

First order tetrahedral elements have four nodes, straight edges, and flat faces. Those edges and faces remain straight and flat after the element has experienced deformation under the applied load. First order tetrahedral elements model the linear field of displacement inside their volume, on faces, and along edges. The linear, or the first order, displacement field gives these elements their name: first order elements. If you recall from the Mechanics of Materials, strain is the first derivative of displacement. Therefore, strain and consequently stress, are both constant in first order tetrahedral elements. This situation imposes a very severe limitation on the capability of mesh constructed with first order elements to model stress distribution of any real complexity. To make matters worse, straight edges and flat faces do not map properly to curvilinear geometry, as shown in figure 1-5.

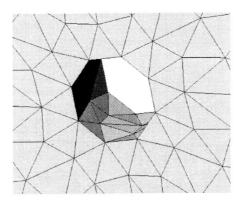

Figure 1-5: Failure of straight edges and flat faces to map to curvilinear geometry

A detail is shown of a mesh created with first order tetrahedral elements. Notice the imprecise element mapping to the hole; flat faces represent the face of the cylindrical hole.

Second order tetrahedral elements have ten nodes that model the second order displacement field in their volume, along faces, and edges. The edges and faces before deformation and after deformation can be curvilinear (second order curve). Therefore, these elements map precisely to curved surfaces, as illustrated in figure 1-6. Second order tetrahedral elements model the second order (parabolic) distribution of displacements inside their volume, on faces, and along edges. Consequently, they model the linear distribution of strains and stresses. Even though they are more computationally demanding than first order elements, second order tetrahedral elements are used for the vast majority of analyses with COSMOSWorks.

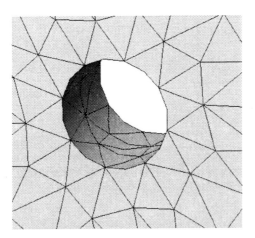

Figure1-6: Mapping curved surfaces

A detail is shown of a mesh created with second order tetrahedral elements. Second order elements map precisely to curvilinear geometry.

Shell elements

Beside solid tetrahedral elements, COSMOSWorks also offers shell elements, used for the meshing of surfaces. While solid elements are created by meshing solid geometry, shell elements are created by meshing surfaces. Shell elements are primarily used for analyzing thin-walled structures. Since surface geometry does not carry information about thickness, the user must provide this information. Similar to solid elements, shell elements also come in draft and high quality with analogical consequences to their ability to map to curvilinear geometry, as shown in figure 1-7 and figure 1-8. As demonstrated with solid elements, first order shell elements model the linear displacement field with constant strain and stress while second order shell elements model the second order (parabolic) displacement field and the first order strain and stress field. Only high quality (second order) shell elements should be used for analysis of any importance.

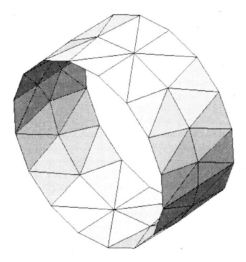

Figure 1-7: First order shell element

This shell element mesh was created with first order elements. Notice the imprecise mapping to curvilinear geometry.

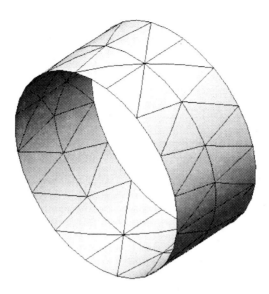

Figure 1-8: Second order shell element

Shell element mesh created with the second order elements, which map precisely to curvilinear geometry.

Certain classes of shapes can be modeled using either solid or shell elements, such as the plate shown in figure 1-9. The type of element, tetrahedral solid or shell, used for modeling depends on the objective of the analysis. More often, however, the nature of geometry dictates what type of element should be used for meshing. For example, parts produced by casting lend themselves to be meshed with solid elements, while a sheet metal structure is best meshed with shell elements.

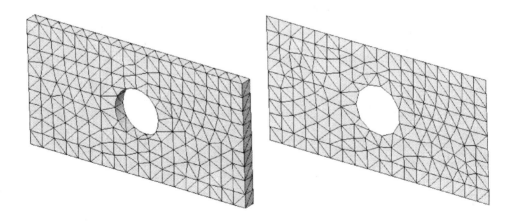

Figure 1-9: Plate modeled with solid elements (left) and shell elements

The plate shown in this illustration can be modeled with either solid (left) or shell elements. The actual choice depends on the particular requirements of analysis and, sometimes, on personal preferences.

Figure 1-10, below, presents the library elements available in COSMOSWorks.

	Triangular shell element 6 Degrees of Freedom per node	Tetrahedral solid element 3 Degrees of Freedom per node
First order elements Linear displacement distribution Constant stress distribution	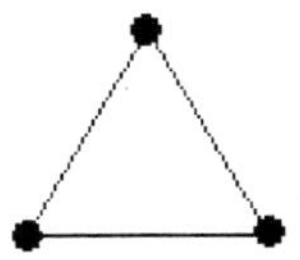	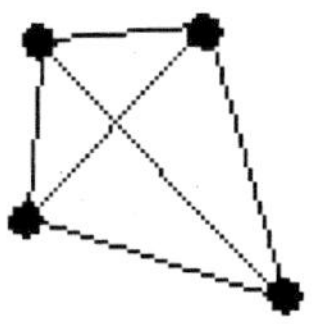
Second order elements Second order displacement distribution Linear stress distribution	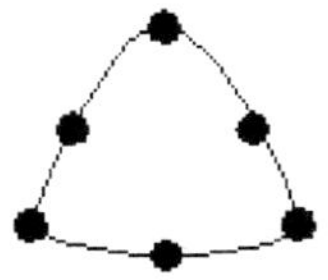	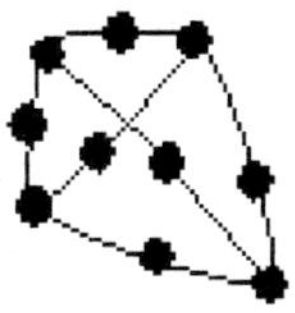

Figure 1-10: COSMOSWorks element library

Four element types are available in the COSMOSWorks element library. The vast majority of analyses use the second order tetrahedral element. Both tetrahedral and shell first order elements should be avoided.

Degrees of freedom

The degrees of freedom (DOF) of a node in a finite element mesh define the ability of the node to perform translation or rotation. The number of degrees of freedom that a node possesses depends on the type of element that the node belongs to. Nodes of solid elements have three degrees of freedom while nodes of shell elements have six degrees of freedom. This means that in order to describe transformation of a solid element from the original to the deformed shape, we need to know only three translational components of nodal displacement of each node, most often the x, y, z displacements. In the case of shell elements, we need to know not only the translational components of nodal displacements, but also the rotational displacement components.

Consequently, built-in (or rigid) constraints applied to solid elements require only three degrees of freedom to be constrained. The same applied to shell element requires that all six degrees of freedom be constrained. Failure to constrain rotational degrees of freedom may result in unintentional hinge support.

What is really calculated in FEA?

Each degree of freedom of each node in a finite element mesh constitutes an unknown. In structural analysis, where we look for deformations and stresses, nodal displacements are primary unknowns. If solid elements are used, there are three displacement components, or three degrees of freedom, per node that must be calculated. Using shell elements, there are six displacement components, or six degrees of freedom, per node that must be calculated when a FEA solution is run. Everything else, such as strains and stresses, is calculated based on the nodal displacements. In fact, some FEA programs offer solutions with stress calculation only as an option.

In thermal analysis, which determines temperatures and heat flow, the primary unknowns are nodal temperatures. Since temperature is a scalar displacement, and not a vector-like displacement, then regardless of what type of elements are used, there is only one unknown (temperature) to be found for each node. All other results available in the thermal analysis are calculated based on nodal temperatures. The fact that there is only one unknown to be found for each node, rather than three or six, makes thermal analysis less computationally intensive than structural analysis.

How do we interpret FEA results?

Results of structural FEA are provided in the form of displacements and stresses. How do we decide between "passed" or "failed" design and what does it take for alarms to go off? What exactly constitutes a failure?

To answer these questions, we need to establish some criteria to interpret FEA results, be they the maximum acceptable deformation, maximum stress, or lowest acceptable natural frequency. While displacement and frequency criteria are quite obvious and easy to establish, stress criteria are not. Let's assume that we conduct a stress analysis in order to ensure that stresses are within an acceptable range. To assess stress results, we need to understand the mechanism of potential failure. If the part breaks, what stress component is responsible for that failure? Is this von Mises stress, the maximum principal stress, shear stress, or yet something else? COSMOSWorks presents stress results in any form we want. It is up to us to decide which stress measure to use for issuing a "pass" or "fail" verdict.

Discussion of various failure criteria would be out of the scope of this book. Interested readers can refer to excellent books, such as those by Spyrakos or Adams and Askenazi, which are listed in chapter 21. Almost any book on the topic of the Mechanics of Materials provides information on this topic. Here we will limit our discussion to outlining the differences between von Mises stresses and the principal stresses used for evaluating material safety.

Von Mises stress

Von Mises stress, also known as Huber stress, is a stress measure that accounts for all six stress components of a general 3-D state of stress (figure 1-11).

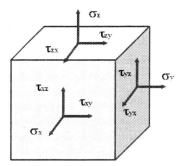

Figure 1-11: Von Mises stress

Two components of shear stress and one component of normal stress act on each side of an elementary cube. Due to equilibrium requirements, the general 3-D state of stress is characterized by six stress components: σ_x, σ_y, σ_z and $\tau_{xy} = \tau_{yx}$, $\tau_{yz} = \tau_{zy}$, $\tau_{xz} = \tau_{zx}$

Von Mises stress σ_{eq}, can be expressed either by stress components that are defined in a global coordinate system as:

$$\sigma_{eq} = \sqrt{0.5*[(\sigma_x - \sigma_y)^2 + (\sigma_y - \sigma_z)^2 + (\sigma_z - \sigma_x)^2] + 3*(\tau_{xy}^2 + \tau_{yz}^2 + \tau_{zx}^2)}$$

or by principal stress, defined as:

$$\sigma_{eq} = \sqrt{0.5*[(\sigma_1 - \sigma_2)^2 + (\sigma_2 - \sigma_3)^2 + (\sigma_3 - \sigma_1)^2]}$$

Note that von Mises stress is a non-negative, scalar value. Von Mises stress is a commonly used stress measure because structural safety for many engineering materials showing elasto-plastic properties (for example, steel) can be described by von Mises stress. The magnitude of von Mises stress is compared to material yield or ultimate strength to calculate the yield strength or the ultimate strength safety factor.

Principal stresses: P1, P2, and P3

The general state of stress can be also represented by three principal stresses: σ_1, σ_2, and σ_3, as shown in figure 1-12. Principal stresses are characterized by the absence of shear stress. In COSMOSWorks, principal stresses are denoted as P1, P2, and P3. The first principal stress, P1, is usually tensile and the third principal stress, P3, is usually compressive. If P3, whose numerical value is lowest of all three stress-components, is compressive, then it becomes the maximum compressive stress.

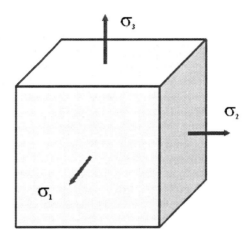

Figure 1-12: General state of stress represented by three principal stresses: σ_1, σ_2, σ_3

P1 stress is used when evaluating stress results in parts made of brittle material, whose safety is better related to P1 than to von Mises stress. P3 is used to examine compressive stresses and contact pressure.

Units of measurements

Internally, COSMOS/Works uses the International System of Units (SI). However, for the user's convenience, the unit manager allows data entry in three different systems of units: SI, Metric, and English. Similarly, results can be displayed any of those three systems of units. Figure 1-13 summarizes the available systems of units and lists some of their measures.

	International System of Units (SI)	**Metric (MKS)**	**English (IPS)**
Mass	kg	kg	lb.
Length	m	cm	in.
Time	s	s	s
Force	N	kgf	lb.
Mass density	kg/m^3	kg/cm^3	$lb./in.^3$
Temperature	°K	°C	°F

Figure 1-13: Unit systems available in COSMOSWorks

SI, metric, and English (IPS) systems of units can be interchanged when entering data in COSMOSWorks.

Experience indicates that units of mass density are often confused with specific gravity. The distinction between those two is quite clear in SI units. Mass density is expressed in kg/m^3 while specific gravity in $[N/m^3]$. In the English system, both specific mass and specific gravity are expressed in $[lb./in.^3]$, where *lb.* denotes either pound mass or pound force.

As COSMOSWorks users, we are spared much confusion and trouble with systems of units; however, we may be asked to prepare data or interpret the results of other FEA software where we do not have the convenience of the unit manager. Therefore, we will make some general comments about the use of different systems of units in the preparation of input data for FEA models. We can use any consistent system of units for FE models, but in practice, the choice of the system of units is dictated by what units are used in the CAD model. The system of units in CAD models is not always consistent; length can be expressed in *mm*, while mass density can be expressed in $[kg/m^3]$. Contrary to CAD models, in FEA all units *must* be consistent. Inconsistencies, which are easy to overlook, especially when defining mass and mass density, lead to very serious errors.

In the SI system, based on meters $[m]$ for length, kilograms $[kg]$ for mass and seconds $[s]$ for time, all other units are easily derived from these basic units: $[m]$, $[kg]$, and $[s]$. The situation gets more complicated if we use a system based on derived units. In mechanical engineering, length is commonly expressed in millimeters $[mm]$, force in Newtons $[N]$, and time in seconds $[s]$.

All other units must then be derived from these basic units: [*mm*], [*N*], and [*s*]. Consequently, the unit of mass is a mass which, when subjected to the unit force on 1 N, will accelerate with the unit acceleration of 1 mm/s^2. Therefore, the unit of mass, in a system using [*mm*] for length and [*N*] for force, is equivalent to 1,000 kg or one metric ton. Consequently, mass density is expressed in metric tonnes [*tonne/mm^3*]. This is critically important to remember when defining material properties in FEA software without a unit manager. Notice in figure 1-14 that erroneous definition of mass density in [*kg/m^3*] rather than in [*tonne/mm^3*] results in mass density being one trillion (10^{12}) times higher (figure 1-14).

System SI	**[m], [N], [s]**
Unit of mass	kg
Unit of mass density	kg/m^3
Density for aluminum	2,794 kg/m^3

System of units derived from SI	**[m], [N], [s]**
Unit of mass	tonne
Unit of mass density	tonne/mm^3
Density of aluminum	2.794 x 10^{-9} tonne/mm^3

English system (IPS)	**[in], [LB], [s]**
Unit of mass	LB = slug/12
Unit of mass density	slug/12/in.^{3}
Density of aluminum	2.614 x 10^{-4} slug/12/in.^{3}

Figure 1-14: Mass densities of aluminum in the three systems of units

Compare mass densities of aluminum defined in the SI system of units with the system of units derived from SI, and with the English (IPS) system of units.

Using on-line Help

COSMOSWorks features very extensive on-line Help and Tutorial functions, which can be accessed from the Help menu in the main COSMOSWorks tool bar (figure 1-15).

Figure 1-15: Accessing the on-line Help and Tutorial

You can access very complete on-line Help and Tutorial functions from the main COSMOSWorks toolbar.

Readers are encouraged to familiarize themselves with the on-line Help function and to supplement the hands-on exercises in this book with information contained in on-line Help, especially when exploring related topics that are not explicitly covered in these exercises.

Limitations of COSMOSWorks

As COSMOSWorks users, we need to appreciate some important limitations of this software: material is assumed as linear, deformations are small, and loads are static. These limitations are not really specific to COSMOSWorks. In fact, they are typical of the vast majority of FEA software used in the design environment.

Linear material

Whatever material we assign to the analyzed parts or assemblies, this material will be assumed as linear, meaning that stress is proportional to strain (figure 1-16).

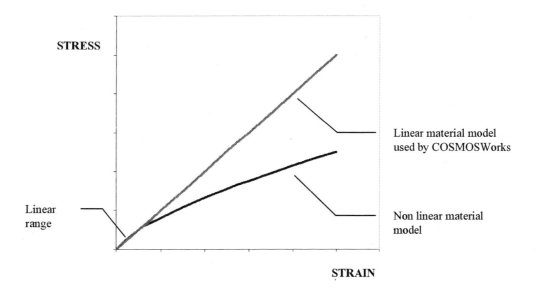

Figure 1-16: Linear material model assumed in COSMOSWorks

In all materials used by COSMOSWorks stress is linearly proportional to strain.

Using a linear material model, the maximum stress magnitude is not limited to yield or to ultimate stress as it is in real life. For example, in a linear model, if stress reaches 800 MPa under a load of 1,000 N, then stress will reach 8,000 MPa under a load of 10,000 N. 8,000 MPa is, of course, a ridiculously high stress value. Material yielding is not modeled, and whether or not yield may in fact be taking place can only be established based on the stress magnitudes reported in results.

Most analyzed structures experience stresses below yield stress, and the factor of safety is most often related to the yield stress. Therefore, the analysis limitations imposed by linear material seldom impede COSMOSWorks users.

Small deformations

Any structure experiences deformation under load. The Small Deformations assumption requires that those deformations be "small". In COSMOSWorks, we assume that those deformations are small. What exactly is a small deformation? Often it is explained as a deformation that is small in relation to the overall size of the structure. For example, small and large deformations of a beam are shown in figure 1-17.

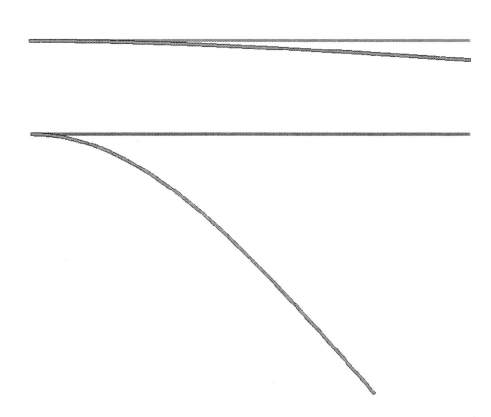

Figure 1-17: Beam in bending with small deformations (top) and large deformations (bottom)

COSMOSWorks software assumes that deformations are small, as those in the top illustration if the beam in bending (undeformed and deformed shapes shown). If deformations are large, as shown in the bottom illustration, COSMOSWorks assumptions do not apply. Other analysis tools, such as COSMOSDesignSTAR, must be used to analyze this structure. COSMOSDesignSTAR is FEA software with nonlinear analysis capabilities that include large deformation analysis.

However, the magnitude of deformation is *not* the deciding factor when classifying deformation as "small" or "large". What really matters is whether or not the deformation changes structural stiffness in a significant way. An analysis run with the assumption of small deformations assumes that the structural stiffness remains the same throughout the deformation process. Large deformation analysis accounts for changes of stiffness caused by deformations. While the distinction between small and large deformations is quite obvious for the beam in figure 1-17, it is not at all obvious for a flat membrane under pressure (figure 1-18).

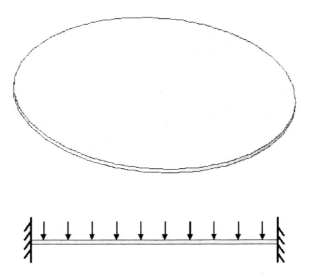

Figure 1-18: Flat membrane (top) under pressure load (bottom)

Here is a classic case where the assumption of small deformation leads to erroneous results. Analysis of a flat membrane under pressure requires a large deformation analysis even though deformations are small in comparison to the size of membrane.

For a flat membrane, initially the only mechanism resisting the pressure load is that of bending stresses. During the deformation process, the membrane acquires membrane stiffness, in addition to the original bending stiffness. Stiffness changes *significantly* during deformation. This change in stiffness requires a large deformation analysis, using tools like COSMOSDesignSTAR.

Static loads

All loads, as well as restraints, are assumed not to change with time, meaning that dynamic loading conditions can not be analyzed with COSMOSWorks. This limitation implies that loads are applied slowly enough to ignore inertial effects.

NOTES:

2: Analysis of a rectangular plate with a hole

Objectives

On completion of this exercise, you will be able to:

- Use the COSMOSWorks interface
- Perform a linear static analysis with solid elements
- Discuss the influence of mesh density on displacement and stress results
- Use different ways to present FEA results

Project description

A steel plate is supported and loaded, as shown in figure 2-1. We assume that the support is rigid (this is also called built-in support) and that the 100,000 N load is uniformly distributed along the end face, opposite to the supported face.

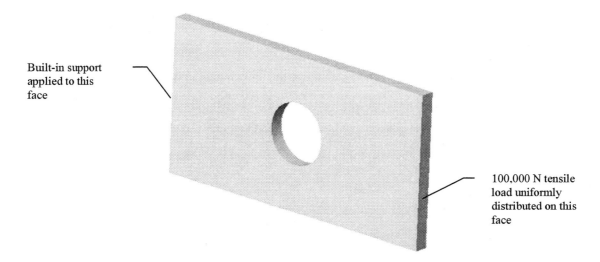

Built-in support applied to this face

100,000 N tensile load uniformly distributed on this face

Figure 2-1: SolidWorks model of a rectangular plate with a hole

We will perform the analysis of displacement and stresses using meshes with different element sizes. Note that repetitive analysis with different meshes does *not* represent standard practice in FEA. We will repeat the analysis using different meshes as a learning tool to gain more insight into how FEA works.

Procedure

In SolidWorks, open the model file called HOLLOW PLATE. Verify that COSMOSWorks is selected in the add-in list. To start COSMOSWorks, select the COSMOSWorks Manager tab, as shown in figure 2-2.

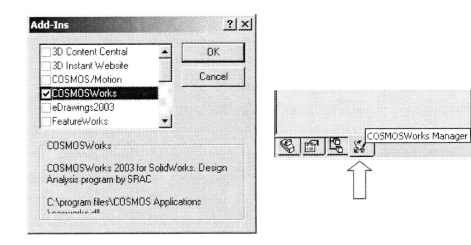

Figure 2-2: Add-Ins list and COSMOSWorks Manager tab

Verify that COSMOSWorks is selected in the list of Add-Ins (left), and then select the COSMOSWorks Manager tab (right).

To create an FEA model, solve the model, and analyze the results, we will use a graphical interface in the form of icons located in the COSMOSWorks Manager window. However, you can achieve the same ends by making the appropriate choices in COSMOSWorks menu. To invoke the menu, select COSMOSWorks from the main tool bar of SolidWorks (figure 2-3).

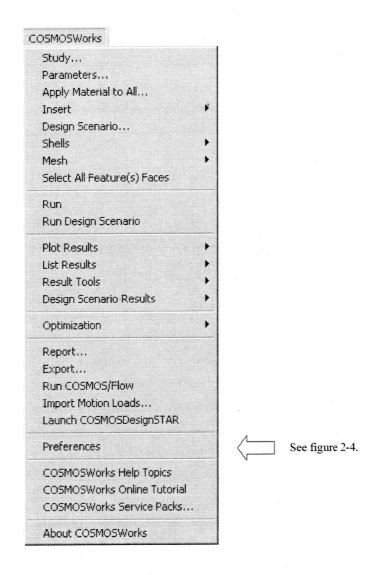

Figure 2-3: COSMOSWorks menu

All functions used for creating, solving, and analyzing a model can be executed either from this menu or from the graphical interface in the COSMOSWorks Manager window. We will use the second method.

Before we create the FEA model, let's review the Preferences window in COSMOSWorks (figure 2-4). You can access the Preferences window from the COSMOSWorks main menu, shown in figure 2-3.

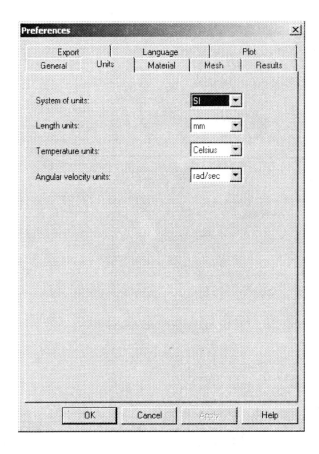

Figure 2-4: COSMOSWorks Preferences window

The COSMOSWorks Preferences window has several tabs. The Units tab, displayed, allows you to define units of measurement. We will use units in the SI system. Please review other tabs before proceeding with the exercise.

Creation of an FEA model always starts with the definition of a study. To define a study, right-click the mouse on the *Part* icon in the COSMOSWorks Manager window and select **Study…**from the pop-up menu. In this exercise, the *Part* icon is called *hollow plate*, as it is in the SolidWorks Manager. Figure 2-5 shows the required selections in the Study definition window: the analysis type is **Static**, the mesh type is **Solid mesh**. Any study name can be used; here we named the study *tensile load*.

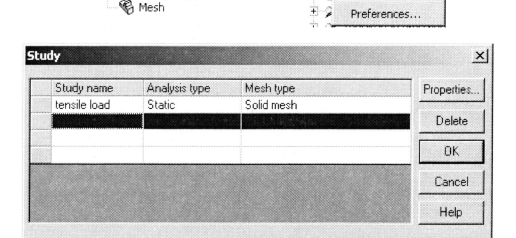

Figure 2-5: Study window

*To display the Study window (bottom), right-click the mouse on the Part icon in the COSMOSWorks Manager window (top left), and then from the pop-up menu, select **Study…**(top right).*

When a study is defined, COSMOSWorks automatically creates a *Study* folder named (in this case) *tensile load* and places several icons in it, as shown in figure 2-6. Notice that some of icons are folders that contain other icons. In this exercise, we will use the *Solids* folder to define and assign material properties and the *Load/Restraint* folder to define loads and restraints. Note that the *Mesh* icon is not part of the *tensile load* folder. If more than one study is defined, they share the same *Mesh* icon.

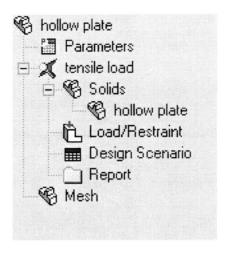

Figure 2-6: *Study* folder

COSMOSWorks automatically creates a Study *folder, called* tensile load, *with the following items:* Solids *folder,* Load/Restraint *folder,* Design Scenario *icon, and* Report *folder. The* Design Scenario *and* Report *folders will not be used in this exercise, nor will the* Parameters *icon, which is automatically created prior to study definition.*

We are now ready to define the mathematical model. This process generally consists of the following steps:

- Geometry preparation
- Material properties assignment
- Restraint (Support) application
- Load application

In this case, the model geometry does not need any preparation (it is very simple as is), so we can start by assigning material properties.

You can assign material properties to the model either by:

- Right-clicking the mouse on the *Solids* folder, or
- Right-clicking the mouse on the *hollow plate* icon, which is located in the *Solids* folder.

However, the first method assigns the *same* material properties to *all* components in the model. The second method assigns material properties to *one particular* component (in this exercise, *hollow plate*). Since we are working with a single part, and not with an assembly, there is no difference between the two methods.

Now, let's right-click the mouse on the *Solids* folder and select **Apply Material to All**. This action opens the Material window shown in figure 2-7.

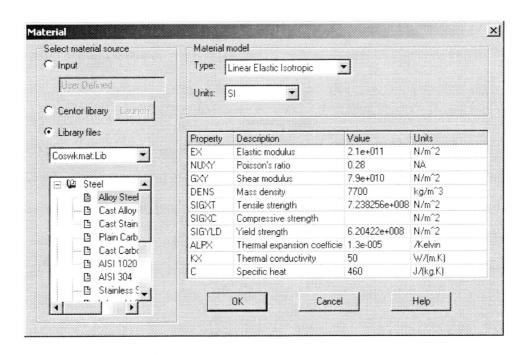

Figure 2-7: Material window

Note: To assign the same materials to all components of the model, right-click the Solids *folder to display the Material window shown above. To assign materials to only one component of the model, right-click the* Part *icon.*

Select **Alloy Steel** in the Material source area, and select **SI** units in the Material model area. Although we use SI units, other units of measurement could be used as well. Notice that the *Solids* folder now shows a check mark and the name of a material to indicate that a material has successfully been assigned. If needed, but not for this exercise, you could define your own material by selecting **Input** in the Select material source area.

Note that material assignment actually consists of two steps:

❑ Material selection or material definition if custom material is used

❑ Material assignment to either all solids in the model or to selected components only (this makes a difference only if the whole assembly is analyzed)

To display a pop-up menu that lists the options available for defining loads and supports, right-click the *Load/Restraint* icon (which will soon become a folder) in the *tensile load* folder (figure 2-8).

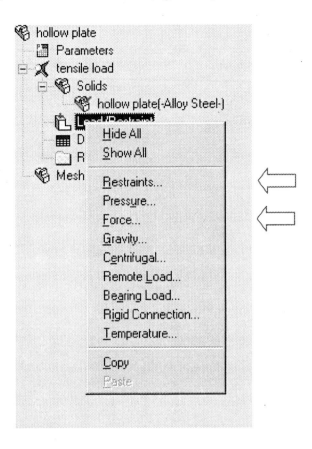

Figure 2-8: Pop-up menu for the *Load/Restraint* folder

The arrows indicate the selections used in this exercise.

To define the restraints that we will use in this exercise, select **Restraints…** from the pop-up menu displayed in figure 2-8. This action opens the Restraint window shown in figure 2-9.

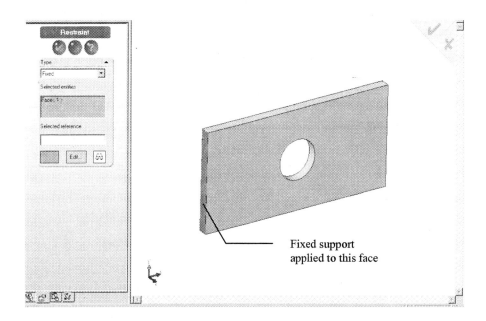

Figure 2-9: Restraint window

The Restraint window also indicates the selected face where fixed restraints are applied.

In this window, you can rotate the model in order to select the face where restraints are applied. Rotate, pan, zoom, and all other view functions work identically as in SolidWorks.

In the Type area, you can select the type of restraint to apply. As you can see in figure 2-9, we requested that fixed restraints be applied to the face. In general, restraints can be applied to faces, edges, and vertexes. To understand the meaning of fixed restraints, we need to review other choices offered in the Restraint window:

Restraint Type	**Definition**
Fixed	Called built-in or rigid support, all translational *and* all rotational degrees of freedom are constrained.
	Note that **Fixed** restraints do not require any information on the direction along which restraints are applied.
Immovable (No translations)	Only translational degrees of freedom are constrained, while rotational degrees of freedom remain unconstrained.
	If solid elements are used, as in this exercise, **Fixed** and **Immovable** restraints have the same effect because solid elements do not have rotational degrees of freedom and only translational degrees of freedom can be constrained.

Restraint Type	Definition
Reference plane or axis	This option restrains a face, edge, or vertex only in a certain direction, while leaving the other directions free to move. You can specify the desired direction of constraint in relation to the selected reference plane or axis.
On flat face	This option provides restraints in selected directions, which are defined by the three principal directions of the flat face where restraints are being applied. **On flat face** offers a very convenient way to apply symmetry boundary conditions, which we will use in later exercises.
On cylindrical face	This option is similar to **On flat face**, except that the three principal directions of a cylindrical reference face define the directions; very useful to apply support that allows for rotation about the axis associated with the cylindrical face.
On spherical face	Similar to **On flat face** and **On cylindrical face**; the three principal directions of a spherical face define the directions of applied restraints.

Having defined restraints, we have fully fixed the model in space. Therefore, the model cannot move without elastic deformation. Any movement of a fully supported model requires deformation. We say that the model does not have any rigid body motions.

Note that the presence of supports in the model is manifested both by the restraint symbols showing on the restrained face and by the automatically created icon, *Restraint-1*, in the *Load/Restraint* folder (which used to display as an icon before the restraints were defined). The display of restraint symbols can be turned on and off either by:

❑ Using the **Hide All** and **Show All** commands in the pop-up menu shown in figure 2-8, or

❑ Right-clicking the restraint symbol for each individually to display a pop-up menu and then selecting **Hide All** from the pop-up menu.

After defining the restraints, we now define loads by selecting **Force** from the pop-up menu shown in figure 2-8. This action opens the Force window, shown in figure 2-10.

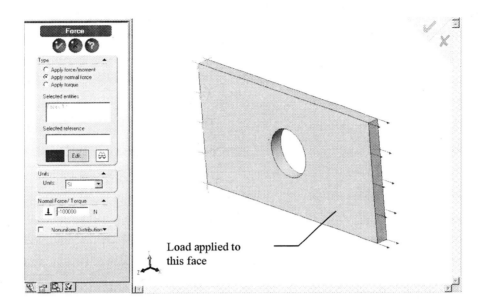

Figure 2-10: Force window

The Force window displays the selected face where normal force will be applied. This illustration also shows symbols of applied restraint and load.

In the Type area, select the **Apply normal force** button in order to load the model with 100,000 N tensile force uniformly distributed over the end face, as shown in figure 2-10. Note that tensile force requires that the force magnitude be defined with a minus sign. Applying positive normal force would result in compressive force.

Generally, forces can be applied to faces, edges, and vertexes using different methods, reviewed below:

Force Type	Definition
Apply force/moment	This option applies force or moment to a face, edge, or vertex in the direction defined by selected reference geometry (plane or axis). You must select the reference geometry before opening the Force window.
	Note that moment can be applied only if shell elements are used. Shell elements have all six degrees of freedom (translations) per node and can take moment load. Solid elements have only three degrees of freedom (translations) per node and, therefore, cannot take moment load directly. If you need to apply moment to solid elements, it must be represented by appropriately distributed forces.
Apply normal force	Available for faces only, this option applies load in the direction normal to the selected face.
Apply torque	Best used for cylindrical faces, this option applies torque about a reference axis using the Right-hand Rule. This option requires that the axis be defined in SolidWorks.

The presence of load(s) is visualized by arrows symbolizing the load and by an automatically created icon, *Force-1*, in the *Load/Restraint* folder.

To familiarize yourself with this feature, right-click the *Restraint-1* and *Force-1* icons to examine the available options. Then use the Click-inside technique to rename those icons. Note that renaming using the Click-inside technique works on all icons in the COSMOSWorks Manager.

We now have built the mathematical model. Before creating the Finite Element model, let's make a few observations about defining:

- Geometry
- Material properties
- Loads
- Restraints

Geometry preparation is a well-defined step with few uncertainties. Geometry that is simplified for analysis can be checked visually by comparing it with the original CAD model.

Material properties are most often selected from the material library and do not account for local defects, surface conditions, etc. Material definition has, therefore, more uncertainties than geometry preparation.

Loads definition, even though done in a few quick menu selections, involves many background assumptions, or guesses, because in real life, load magnitude, distribution, and time dependence are often known only approximately and must be roughly estimated in FEA with many simplifying assumptions. Therefore, significant idealization errors can be made when defining loads. Still, loads can be expressed in numbers and the load symbols provide visual feedback on load direction, making loads easier for the FEA user to relate to.

Defining restraints is where severe errors are most often made. It is easy enough, for example, to apply a built-in restraint without giving too much though to the fact than built-in restraint means a rigid support, which is a mathematical abstract. A common error is over-constraining the model, which results in an overly stiff structure that will underestimate deformations and stresses. The relative level of uncertainties in defining geometry, material, loads and restraints is qualitatively shown in figure 2-11.

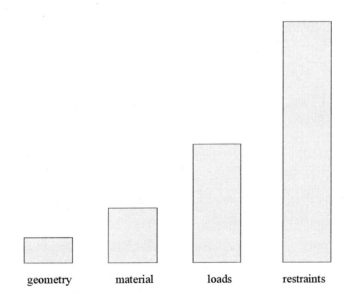

Figure 2-11: Qualitative assessment of relative levels of difficulty and uncertainties in defining geometry, material, loads, and restraints

Geometry is easiest to define while restraints are the most difficult.

The level of difficulty has no relation to time required for each step, so the message in figure 2-11 may be counterintuitive. In fact, preparing CAD geometry for FEA may take hours, while applying restraints takes only a few mouse clicks.

In all examples here, we will assume that material properties, loads, and supports are known, *with certainty*, and that the way they are defined in the model represents an acceptable idealization of real conditions. However, we need to point out that it is the responsibility of user of the FEA software to determine if all those idealized assumptions made during the creation of the mathematical model are indeed acceptable. Even the best automesher and the fastest solver will not help if the mathematical model submitted for FEA is based on erroneous assumptions.

To open the pop-up menu for meshing, shown in figure 2-12 right, right-click the *Mesh* icon (figure 21-12 left).

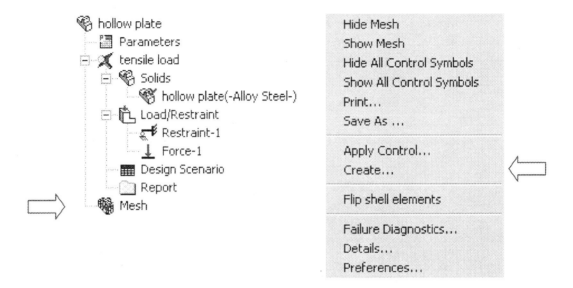

Figure 2-12: *Mesh* icon (left) and the pop-up menu for meshing (right)

In the pop-up menu, select **Create...** to open the Mesh window. This window offers a choice of element size and element size tolerance. In this exercise, we wish to study the impact of mesh size on results, so we will solve the same problems using three different meshes: coarse, medium (default), and fine. We will create these meshes using different selections of meshing parameters, as shown in figure 2-13.

It is important to point out that this activity is *not* the standard practice with COSMOSWorks, or any other FEA tool for that matter. We will use three different meshes and solve the same problem three times only as a part of the learning process.

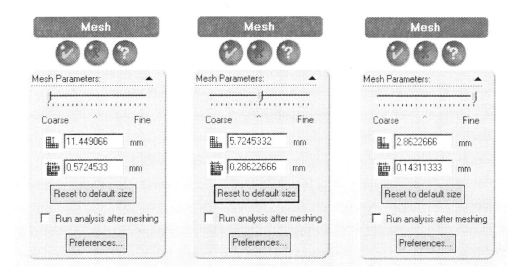

Figure 2-13: Three choices for mesh density: coarse (left), medium (center), and fine (right)

Medium density is the default choice offered by the COSMOSWorks mesher.

The medium mesh density, shown in the middle window in figure 2-12, is default that COSMOSWorks proposes for meshing our tensile strip model. The element size of 5.72 mm and the element size tolerance of 0.286 are established automatically based on the geometric features of the SolidWorks model. The 5.72-mm size is the characteristic element size in the mesh, as explained in figure 2-14. The element size tolerance is the allowable spread of the actual element sizes in the mesh.

Mesh density has a direct impact on the accuracy of results. The smaller the elements, the lower the discretization errors, but meshing and solution take longer. In the majority of analyses with COSMOSWorks, the default mesh settings produce meshes that provide acceptable discretization errors while keeping solution times reasonably short.

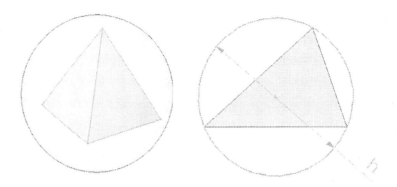

Figure 2-14: Characteristic element size for a tetrahedral element

The characteristic element size for a tetrahedral element is the diameter of a circumscribed sphere (left). This is easier to illustrate with the 2-D analogy of a circle circumscribed on a triangle (right).

The characteristic element size for 2-D triangular element (not available in COSMOSWorks) is the diameter (h) of a circle circumscribed on the triangular element. For 3-D tets the characteristic element size is the diameter of a sphere circumscribed on a tetrahedron.

The characteristic element size for 2-D element is the diameter (h) of a circle circumscribed on the triangular element. For 3-D elements the characteristic element size is the diameter of a sphere circumscribed on a tetrahedron.

Before we proceed with meshing, we need to open the Preferences window, which can be displayed from meshing pop-up menu (figure 2-11). Click the Mesh tab. The Mesh tab in the Preferences window is shown in figure 2-15. We want to ensure that the mesh quality is set to **High**. The difference between **High** and **Draft** mesh quality is that:

❑ Draft quality mesh uses first order elements

❑ High quality mesh uses second order elements

We discussed the differences between first and second order elements in chapter 1.

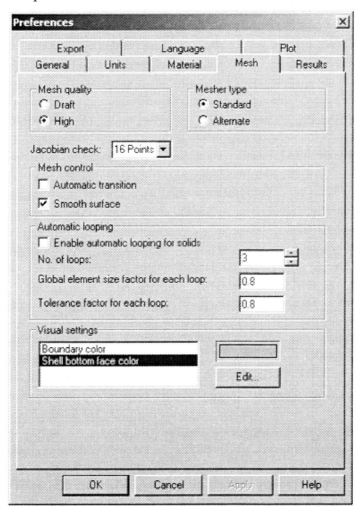

Figure 2-15: Mesh tab in the Preferences window

We use this window to verify that the choice of mesh quality is set to High.

Having verified that high quality mesh is selected, close the Preferences window, and then right-click the *Mesh* icon again and select **Create...** as shown in figure 2-12. This opens Mesh window, the one with the slider.

With the Mesh window open, set the slider all the way to the left (as illustrated in figure 2-13) and select (check) the Run analysis after meshing checkbox to create a coarse mesh. The mesh will be displayed as shown in figure 2-16.

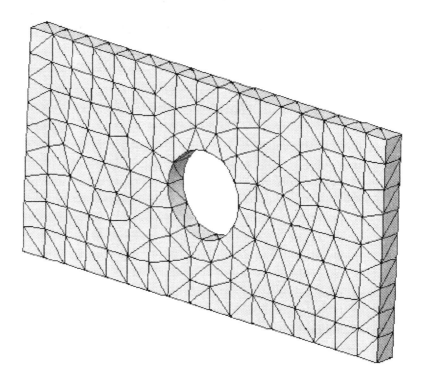

Figure 2-16: Coarse mesh created with second order, solid tetrahedral elements

You can control the mesh visibility by selecting **Hide Mesh** *or* **Show Mesh** *from the pop-up menu shown in figure 2-12.*

We are now ready for running the solution.

To start the solution, right-click the *Study* icon, here renamed *tensile load*. This action displays a pop-up menu (figure 2-17).

Figure 2-17: Pop-up menu for the *Study* icon

Start the solution by right-clicking the Study *icon to display a pop-up menu. Select* **Run** *to run the solution.*

Select **Run** from the pop-up menu to run the solution. The solution can be executed with different properties, which we will investigate in later chapters. You can monitor the solution progress in a window while the solution is running (figure 2-18).

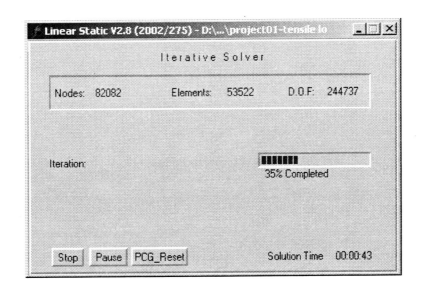

Figure 2-18: Solution Progress window

The Solver reports solution progress while the solution is running.

Successful completion of solution or a failed solution is reported, as shown in figure 2-19 and must be acknowledged before proceeding.

Figure 2-19: Solution outcome: completed or failed

Once the solution is completed, COSMOSWorks automatically creates several new folders in the COSMOSWorks Manager window:

- Stress
- Displacement
- Strain
- Deformation
- Design Check

Each folder holds an automatically created plot with its respective type of result (figure 2-20). The Stress, Displacement, Strain, Deformation and Design Check plots are ready for examination. The Design Check plot requires user input before viewing. If desired, you can add more plots to each folder.

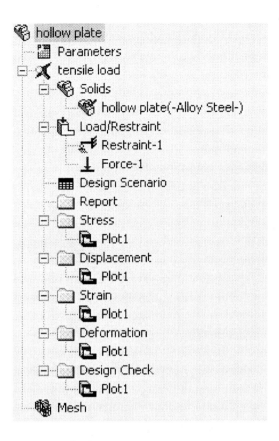

Figure 2-20: Automatically created *Results* folders

One default plot of respective results is contained in each of the automatically created Results *folders: Stress, Displacement, Strain, Deformation, and Design Check.*

You can modify plots by right-clicking the *Plot* icon. If desired, you can add new plots by right-clicking the respective *Results* folder.

Prior to reviewing stress results, it is worthwhile to examine the Stress plot window displayed by right-clicking the *Plot1* icon in the *Stress* folder, which opens a pop-up menu, and then selecting **Edit definition...** The Stress plot window (figure 2-21, right) has several tabs. The contents of the Display tab are shown in figure 2-21. These options define the type of stress component shown in the plot (here **von Mises stress**) and the type of graphics display used (here **Filled, Discrete**). Please investigate the Properties and Settings tabs before you proceed.

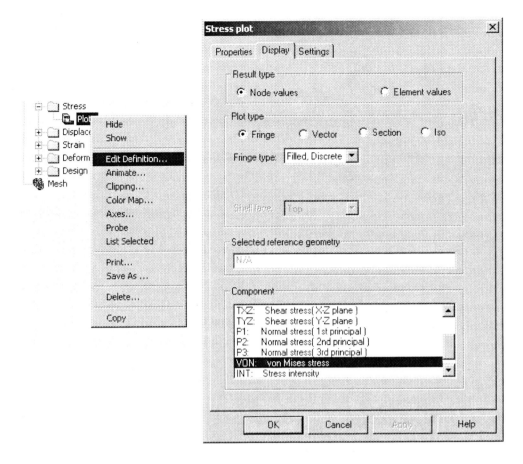

Figure 2-21: Stress plot window

The Stress plot window defines how stress results are displayed. The contents of the Display tab are shown.

Please investigate what Properties and Settings tabs have to offer. In particular examine different units options under the Properties tab. Also examine the options shown in the pop-up window on the left.

We will now review the stress, displacement, strain, and deformation results. All of these plots are created and modified in the same way. Sample results are shown in:

❑ Figure 2-22 (von Mises stress)

❑ Figure 2-23 (displacement)

❑ Figure 2-24 (strain)

❑ Figure 2-25 (deformation)

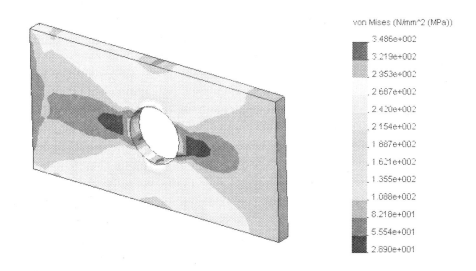

Figure 2-22: Von Mises stress results using **Filled, Discrete** display

*In the Von Mises stress results using the **Filled, Discrete** display, notice the display units (MPa).*

If desired, you can change the display units in the Stress plot window under the Properties tab. Stresses are presented here as nodal stresses, also called averaged stresses. Elements (or non-averaged stresses) can also be displayed by modifying the settings in the Stress Plot window under the Settings tab. Nodal stresses are most often used to present stress results. See chapter 3 and the glossary of terms at the end of this book for more comments on nodal and element stresses.

Figure 2-23 shows the displacement results.

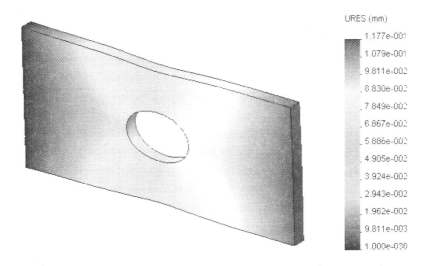

Figure 2-23: Displacement results using **Filled, Gouraud** display

*In the displacement results using **Filled, Gouraud** display, notice the display units (mm).*

This plot shows the deformed shape in an exaggerated scale. You can change the display from undeformed to deformed and modify the scale of deformation in the Stress Plot window under the Settings tab.

Figure 2-24 shows the strain results.

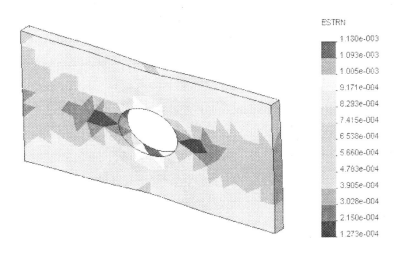

Figure 2-24: Strain results

Notice that strain is dimensionless. As opposed to stress results, which are commonly shown as averaged (nodal stresses), strain results are always shown as non-averaged.

Figure 2-25 shows the deformation results.

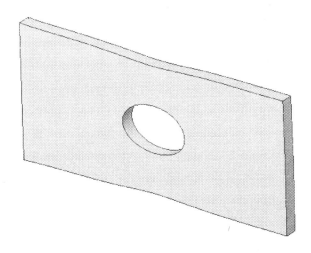

Figure 2-25: Deformation results

Notice that deformed plots can be also created in all previous types of display if the deformed shape display is selected.

Deformation is shown in an exaggerated scale. You can modify the scale in the Stress plot window under the Settings tab.

The last folder, called *Design Check* (figure 2-20), holds the *Plot1* icon, which was automatically placed in it; however, user input is required before the plot can be viewed. To view the plot, double-click the *Plot1* icon. This action displays the first of three windows in the *Design Check* wizard (shown in figure 2-26). Subsequent windows are shown and explained in figure 2-27 and figure 2-28.

Figure 2-26: First of three windows in the Design Check wizard

Use this window to select the design evaluation criterion.

We decided to base the design evaluation on von Mises stress. We decide to base the design evaluation on von Mises stresses. In other words, we think that structural safety can be adequately expressed by von Mises stress.

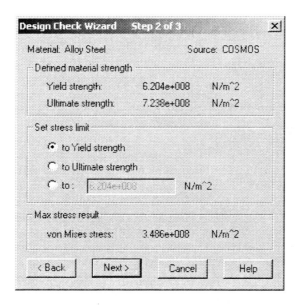

Figure 2-27: Second of three windows in the Design Check wizard

Use this window to select the stress limit used for calculating the factor of safety.

The values for yield strength and ultimate strength (tensile strength) come from material definition shown in figure 2-7. If desired, you can use a custom value. Our choice here is **Yield strength**.

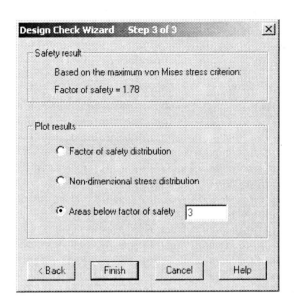

Figure 2-28: Third of three windows in the Design Check wizard

Use this window to determine how to plot the results.

Here we determined that areas with a safety factor below 3 be shown in red. The Design Check plot is shown in figure 2-29.

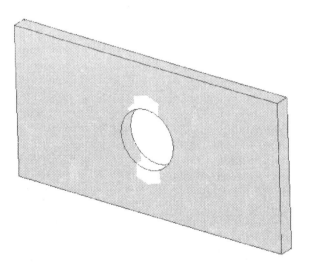

Figure 2-29: Red color (shown as light gray in this grayscale illustration) displays areas at risk

The red color indicates areas where the factor of safety is below 3. Only one type of display (Filled, Gouraud) is available.

We have completed the analysis with a coarse mesh and now wish to see how a change in mesh density will affect the results. Therefore, we will repeat the analysis two more times using medium and fine density meshes respectively. We will use the settings shown in figure 2-13. All three meshes used in this exercise (coarse, medium, and fine) are shown in figure 2-30.

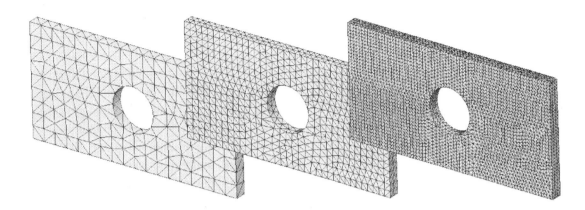

Figure 2-30: Models with coarse, medium, and fine mesh densities for comparison of results

We will use three meshes to study the effect of mesh density on the results of analysis.

To compare the results produced by different meshes, we need more information than is found in the Results plots. Along with the maximum displacement and the maximum von Mises stress, for each mesh, we need to know:

❑ Number of nodes

❑ Number of elements

❑ Number of degrees of freedom

This information on the number of nodes and number of elements can be found in Mesh Details (figure 2-31). The number of degrees of freedom in the model is equal to the number of nodes times the number of degrees of freedom per node (in our case three degrees of freedom per node) not counting the nodes on the face where support is defined.

We need to write down results of each step of mesh refinement because after remeshing previous results are no longer available (figure 2-32).

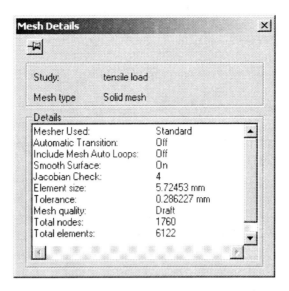

Figure 2-31: Meshing details window

Right click on Mesh icon and select Details… to activate this window.

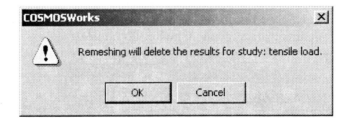

Figure 2-32: COSMOSWorks dialog box

Note that re-meshing deletes the current results.

The summary of results produced by the three models is shown in figure 2-33. We need to stress that all the results of this exercise pertain to the same problem. The only difference is in the mesh density.

mesh density	max. displ. magnitude	max. von Mises stress	number of DOF	number of elements	number of nodes
coarse	0.1177 mm	348.6 MPa	6684	1109	2285
medium	0.1180 mm	368.1 MPa	32439	6122	10998
fine	0.1181 mm	375.9 MPa	244737	53522	82082

Figure 2-33: Summary of results produced by the three meshes

Note that these results are based on the same problem. Differences in the results arise from the different mesh densities used.

Figures 2-34 and 2-35 show the maximum displacement and the maximum von Mises stress as functions of the number of degrees of freedom, which corresponds directly to the mesh density.

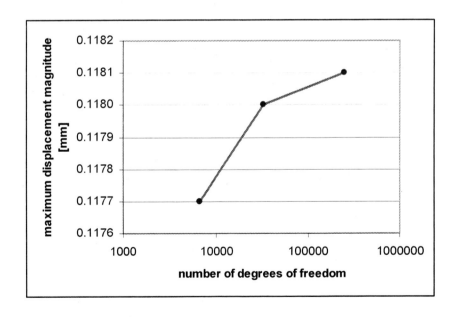

Figure 2-34: Maximum displacement magnitude

Maximum displacement magnitude is plotted as a function of the number of degrees of freedom in the mode. The three points on the curve correspond to the three models solved. Straight lines connect the three points only to visually enhance the graph.

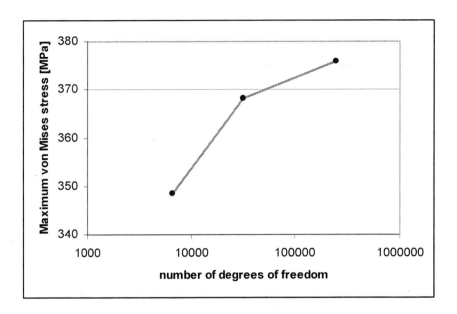

Figure 2-35: Maximum von Mises stress

Maximum von Mises stress is plotted as a function of the number of degrees of freedom in the model. The three points on the curve correspond to three models solved. Straight lines connect the three points only to visually enhance the graph.

Having noticed that the maximum displacement increases with mesh refinement, we can conclude that the model becomes "softer" when smaller elements are used. This result stems from the artificial constrains imposed by element definition becoming less imposing with mesh refinement. In our case, by selecting second order elements, we imposed the assumption that the displacement field in each element is described by the second order polynomial function. While with mesh refinement, the displacement field in each element remains the second order polynomial function, the larger number of elements makes it possible to approximate the real displacement and stress field more accurately. Hence, we can say that the artificial constraints imposed by element definition become less imposing with mesh refinement.

Displacements are always the primary unknowns in FEA, and stresses are calculated based on displacement results. Therefore, stresses also increase with mesh refinement. If we continued with mesh refinement, we would see both displacement and stress results converge to a finite value. This limit is the solution of the mathematical model. Differences between the solution of the FEA model and the mathematical model are due to discretization error. Discretization error diminishes with mesh refinement.

We will now repeat our analysis of the hollow plate using prescribed displacements in the place of load. Prescribed displacement is an alternate way of loading the model. Rather than loading it with a 100,000 N force that causes 0.118-mm displacement of the right end-face, we will apply the prescribed displacement of 0.118 mm to this face to see what stresses this causes. For this exercise, we will use only one mesh with default (medium) mesh density.

In order to have both sets of results available for analysis and comparison, we will define the second study, called ***tensile load pd***. The definition of material properties and of the built-in support to the left-side end-face is identical to the previous design study.

For the new study, we can either:

- Repeat the materials definition and assignment, or
- Drag the *hollow plate* icon from the *Solids* folder (in the *tensile load* study) and drop it into the *Solids* folder in the *tensile load pd* folder.

Similarly we can drag and drop the *Restraint-1* icon from the *tensile load* study into *Load/Restraint* folder in the *tensile load pd* study.

To apply the prescribed displacement to the right-side end-face, we need to select this face and define the prescribed displacement as shown in figure 2-36. The minus sign is necessary to obtain displacement in the tensile direction.

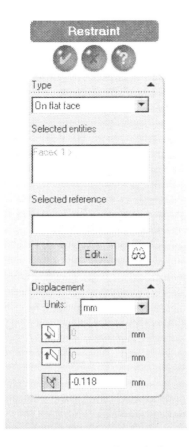

Figure 2-36: Restraint window

The prescribed displacement of 0.118 mm is applied to the same face where the tensile load of 100,000 N had been applied.

Notice that prescribed displacement overrides the force load if that is still applied. While it is better to delete the force load in order to keep the model clean, the force load has no effect if prescribed displacements are applied to the same entity.

We now need to mesh the model with the default mesh density, re-run the *tensile load* study, and run the *tensile load pd* study. We need to re-run *tensile load* study because it was last run with a high mesh density. Now we want to have the results of both studies produced by the same mesh, with default element size of 5.72 mm. Figure 2-37 shows both studies solved sharing the same mesh.

Engineering Analysis with COSMOSWorks

Figure 2-37: COSMOSWorks manager window showing both studies solved

Defining and solving the two studies in one model allow for comparison of their results.

Figures 2-38 and 2-39 compare displacement and stress results for both studies.

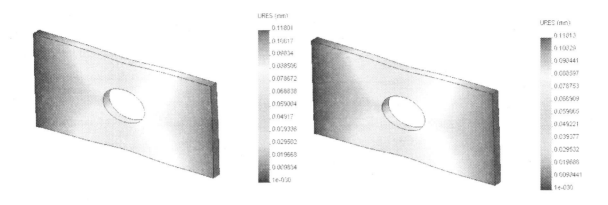

Figure 2-38: Comparison of displacement results

Displacement results in the model with the force load are displayed on the left and displacement results in the model with the prescribed displacement load are displayed on the right.

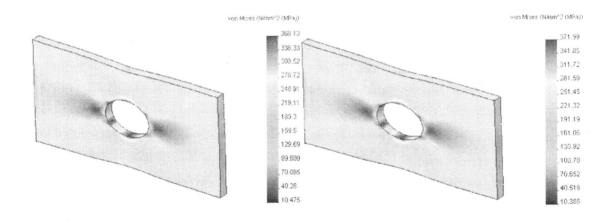

Figure 2-39: Von Mises stress results

Von Mises stress results in the model with the force load are displayed on the left and Von Mises stress results in the model with the prescribed displacement load are displayed on the right.

Note different numerical format of results. You can change the format in the Preferences window (figure 2-4) under the Plot tab.

Note that the results produced by applying force load and by applying prescribed displacement load are very close, but not identical. The reason for this discrepancy is that in the model with the force load, the loaded face is allowed to deform. In the model loaded with prescribed displacement, this face remains flat, even though it experiences displacement as a whole. Also, the prescribed displacement of 0.118 mm applies to the entire face in the model with prescribed displacement, while it is the maximum displacement for only some points on the face in the model with the force load.

We will conclude our analysis of the hollow plate by examining the reaction forces. If any result plot is still displayed, hide it now (right-click the *Plot* icon and select **Hide** from pop-up window). Before we access the reaction force results, we first need to select the face for which we wish to obtain the reaction force results. In this case, it is the face where the built-in support was applied. Having selected the face, right-click the *Displacement* folder. A pop-up menu appears (figure 2-40). Select **Reaction Force....**

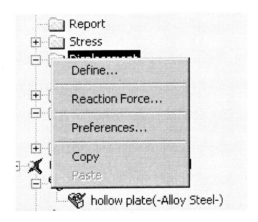

Figure 2-40: Pop-up menu associated with the *Displacement* folder

Right-click the Displacement *folder to display a pop-up menu that allows you to open the Reaction Force window.*

Figure 2-41 shows Reaction Force results for both studies: with force load (left) and with prescribed displacement load (right).

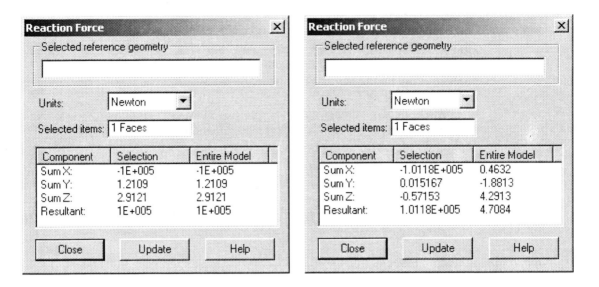

Figure 2-41: Comparison of reaction force results

Reaction forces are shown on the face where built-in support is defined for the model with force load (left) and for the model with prescribed displacement load (right).

3: Analysis of an L-bracket

Objectives

On completion of this exercise, you will be able to:

- Illustrate the differences between modeling and discretization errors
- Use mesh controls
- Describe when the lack of convergence of FEA results may occur

Project description

An L-shaped bracket (in a file called L BRACKET in SolidWorks) is supported and loaded as shown in figure 3-1. We wish to find the displacements and stresses caused by a 1,000 N bending load. In particular, we are interested in stresses in the corner where the 2-mm round is located. Since the radius of the round is small compared to the overall size of the model, we decide to suppress it. As we will soon prove, suppressing the round is a bad mistake!

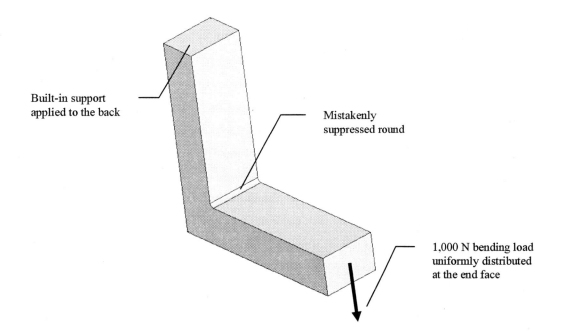

Figure 3-1: Loads and supports applied to an L-shaped bracket

Notice that the round has been suppressed, leaving in its place a sharp re-entrant corner.

Procedure

Following the same steps as those described in chapter 2, we define the study, assign material properties (use Alloy Steel), loads, and supports. Next we mesh the model with second order tetrahedral elements, accepting the default settings of the automesher. The finite element mesh, which we will call *mesh 1*, is shown in figure 3-2.

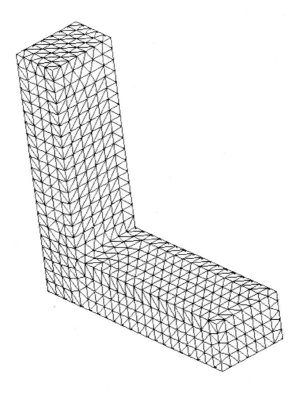

Figure 3-2: Finite element mesh using the default setting in the automesher

In this finite element mesh, the element size is 4.76 mm.

The displacement and stress results obtained using *mesh 1* are shown in figure 3-3.

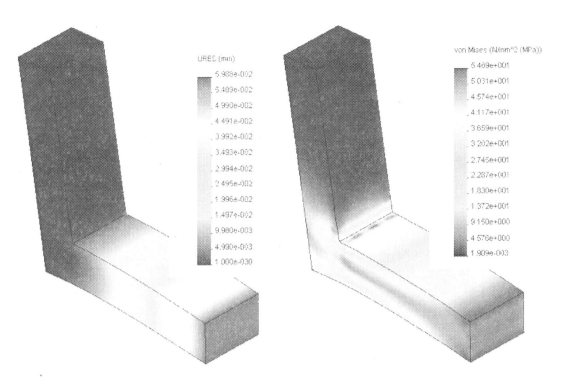

Figure 3-3: Displacements and von Mises stress results produced using *mesh 1*

The maximum displacement is 0.06 mm; the maximum von Mises stress is 54.9 MPa.

Now we will investigate the effect of using smaller elements on the results. In chapter 2, we did this by refining the mesh uniformly so that the entire volume of the model was controlled by the global element size.

In this exercise, we will use a different technique. Having noticed stress concentration near the sharp re-entrant corner, we will refine the mesh locally in that area by applying mesh controls.

First select the edge where mesh controls will be applied. Then right-click the *Mesh* icon to display the pop-up menu shown in figure 3-4.

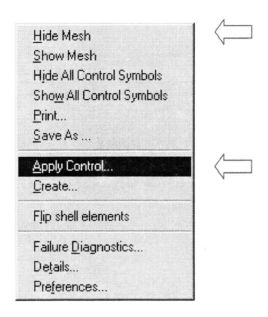

Figure 3-4: Pop-up menu

Using the pop-up menu, we first hide the existing mesh, and then select **Apply Control…**, which opens the Mesh Control window (figure 3-5). Note that it is also possible to open the Mesh control window first and then select the desired entity or entities (here the re-entrant edge) where mesh controls are to be applied.

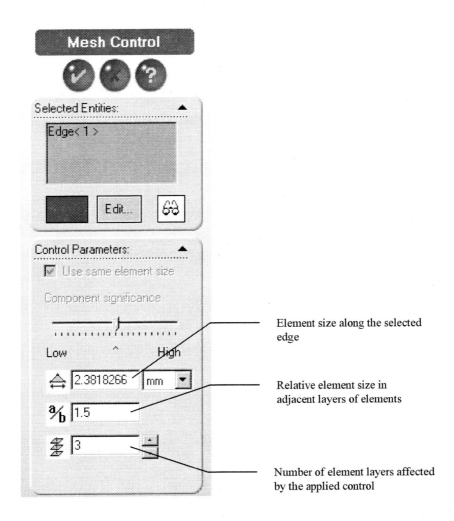

Figure 3-5: Mesh Control window

Mesh controls allow you to control the local element size on selected entities.

The element size along the selected edge can now be controlled independently of the global element size. Now we can create a second mesh with the same global element size, 4.76 mm, but the element created along the selected edge will be 2.38 mm. Mesh control can also be applied to vertexes, faces, or entire components of assemblies. Notice that once mesh controls have been defined, the *Mesh* icon becomes a folder. Mesh controls can then be edited using a pop-up menu displayed by right-clicking the *Control-1* icon, which is now located in the *Mesh* folder (figure 3-6).

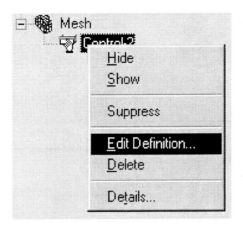

Figure 3-6: Pop-up menu for the *Mesh* icon

The Mesh icon becomes a folder after mesh controls have been defined. Mesh control can be manipulated using the pop-up menu displayed by right-clicking the Control-2 icon in the Mesh folder.

Note that now a Control-2 icon, not the Control-1 is showing. This change is because, in the process of preparing this exercise, the author has deleted and re-applied the mesh control.

The mesh with applied control (also called mesh bias) is shown in figure 3-7.

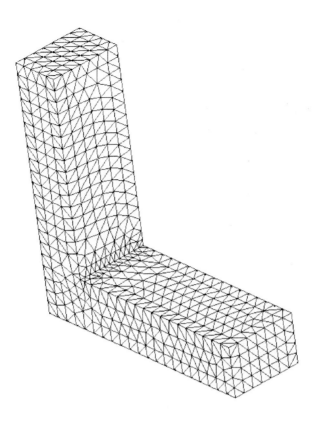

Figure 3-7: Mesh with applied controls (mesh bias)

Mesh 2 is refined along the selected edge. The effect of mesh bias extends for three layers of elements adjacent to the edge.

Maximum displacements and stress results obtained using mesh 2 *are 0.06011 mm and 62.46 MPa respectively.*

Now we repeat the same exercise once again with a different mesh control. To do this, we can either:

❑ Edit the existing mesh control (figure 3-8), or

❑ Delete the existing mesh control and define a new one.

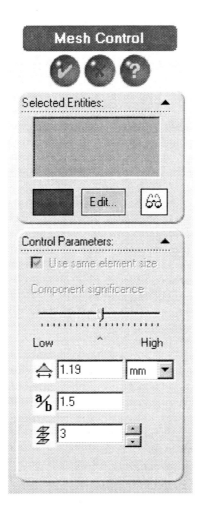

Figure 3-8: Mesh Control window

The element size along the edge is now 1.19 mm, half of what was shown in figure 3-5. Note that nothing yet appears in the Selected entities field. This is because you can select the edge can be after you open the Mesh Control window. This functionality also often applies to loads and restraints.

Figure 3-9 shows *Mesh 3*, which is created using the mesh bias defined in figure 3-8.

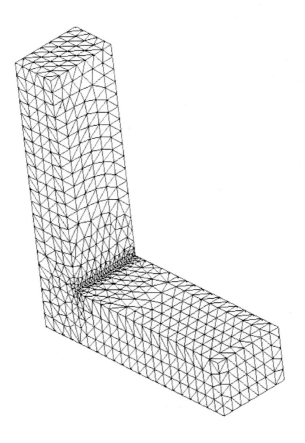

Figure 3-9: *Mesh 3*, created with the mesh control shown in figure 3-8

Maximum displacements and stress results obtained using Mesh 3 *are 0.0631 mm and 84.00 MPa.*

The results of all three runs are summarized in figure 3-10.

Mesh	Maximum Displacement (*mm*)	% Increase	Maximum Stress (*MPa*)	% Increase
1	0.05988		54.89	
2	0.06011	0.38	62.46	14
3	0.06031	0.33	84.00	34

Figure 3-10: Summary of displacement and stress results produced by meshes

Note that each mesh refinement brought about an increase in both the maximum displacement and the maximum stress. The increase in the displacement result is negligible and decreases with successive runs.

Hypothetically, we could continue with this exercise of progressive mesh refinement, either:

❑ Locally, near the sharp re-entrant, as we have done here by means of mesh controls, or

❑ Globally, by reducing the global element size, as we did in chapter 2.

If so, we would notice that displacement results converge to a finite value and that even the first mesh is good enough if we are looking only for displacements.

Stress results, however, behave quite differently. Each subsequent mesh refinement produces higher stress results. Instead of converging to a finite value, the stress results diverge. This divergence is illustrated in figure 3-11, which shows the maximum von Mises stress results obtained with five different meshes (the author expanded this exercise by creating two more meshes).

Given enough time and patience, we can produce results showing any stress magnitude we want. All that is necessary is to make the element size small enough!

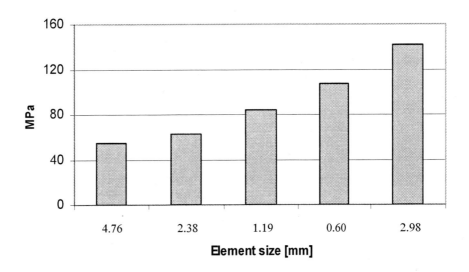

Figure 3-11: Maximum von Mises stress results produced by five increasingly refined meshes

These maximum von Mises stress results were produced by five meshes with different mesh controls that affected the element size along the selected edge of the model.

The reason for divergent stress results is not that the finite element model is incorrect, but that that the finite element model is based on the wrong mathematical model.

According to the theory of elasticity, stress in the sharp re-entrant corner is infinite. The finite element model does not produce infinite stress results due to discretization errors, and these discretization errors simply mask the modeling error. However, stress results are completely dependent on mesh size; therefore, they are totally meaningless.

If our objective is to find the maximum stress, then the decision to suppress the round and analyze a model with a sharp re-entrant corner is a very serious mistake. The maximum stress in a sharp re-entrant corner is singular, or infinite. The round, no matter how small it is, must be included in the model if we need to find stresses in or near that round.

Therefore, we must repeat this exercise using a model with the round. Obtaining the correct model requires unsuppressing the round, which is done in the SolidWorks Manager.

Notice that after the round has been unsuppressed and we return from the SolidWorks Manager window to the COSMOSWorks window, COSMOSWorks displays a message that the geometry has changed and the model requires remeshing and rerunning. This well illustrates the connectivity between SolidWorks and COSMOSWorks. COSMOSWorks also keeps track of the validity of loads and restraints. If any of these become invalid as a result of geometry modifications performed by SolidWorks, COSMOSWorks issues appropriate warnings.

Because the round is a small feature compared to the overall size of the model, meshing with the default mesh settings will produce an abrupt change in element size between the round and adjacent faces. Also, the elements modeling the round may be insufficient. To avoid this problem, we need to select (check) the Automatic transition option in the Preferences window under the Mesh tab (figure 3-12).

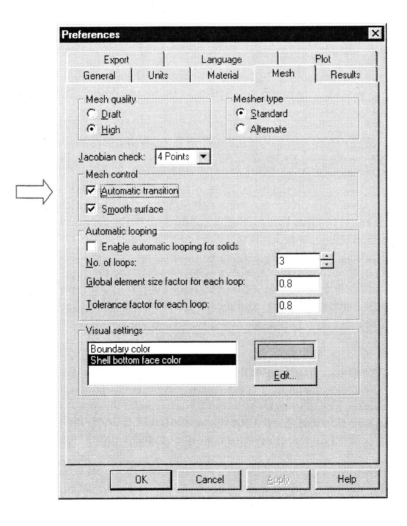

Figure 3-12: Meshing preferences with **Automatic transition** selected

Figure 3-13 shows von Mises stress results superimposed on the plot of the finite element mesh, with and without the **Automatic transition** option applied. You can display the mesh together with any results by modifying the options in the Stress plot window under the Settings tab as shown in figure 3-14.

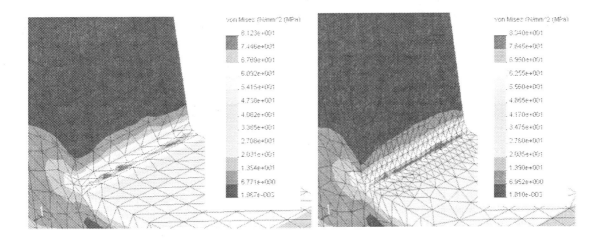

Figure 3-13: Von Mises stress results in a model with a round

*Compare meshes created with (right) and without (left) the **Automatic transition** option.*

Engineering Analysis with COSMOSWorks

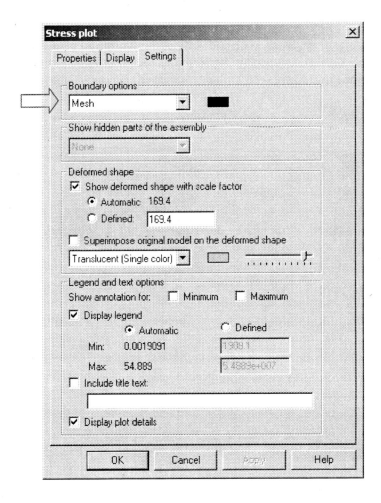

Figure 3-14: Settings for displaying the mesh together with the results

*Displaying the mesh together with the results requires selecting **Mesh** in the Boundary options area in the Stress plot window under the Settings tab.*

The L-BRACKET example is a good place to review the different ways of displaying stress results. Figure 3-15 shows the node values of von Mises stress produced by the first mesh we used in this exercise, the one with default element size and no mesh control. Figure 3-16 shows element values. As we can see, these two plots are quite different (Node or Element stress display is selected in the Stress plot window under the in Display tab).

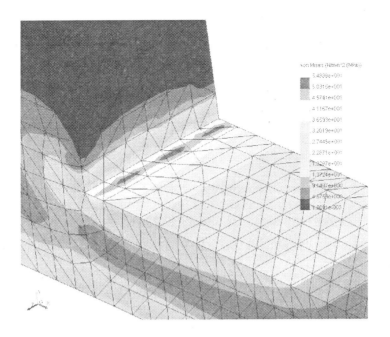

Figure 3-15: Von Mises stresses displayed as node values

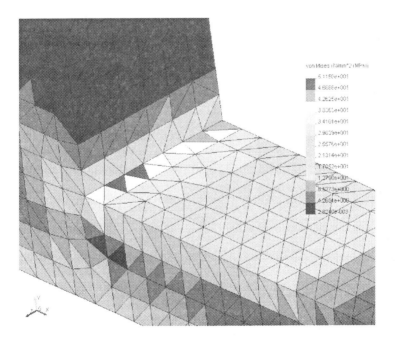

Figure 3-16: Von Mises stresses displayed as element values

To understand what node and element values are, why they are different, and how can we use those different display options, we need to explain how stresses are calculated.

As we explained in chapter 1, nodal displacements are computed first. Strains and then stresses are calculated after displacements have been found. Stresses are first calculated inside the element in certain points, called Gaussian points. Next, stress results are extrapolated to element nodes. This extrapolation is done for all elements. If one node belongs to more than one element, as is always the case except for vertex nodes, then the stress results from all those elements sharing a given node are averaged and one stress value, called a node value, is reported for each node.

An alternate procedure to present stress results can also be used. Having obtained stresses in Gaussian points, those stresses are extrapolated to nodes just like before, but they are not averaged with stresses from the other elements. Instead, they are averaged among themselves. This means that one stress value is calculated for the element by averaging the stress results on all nodes of the same element. This value is called an element value. The reason why node values and element values differ is because the neighboring elements have different stresses.

Node values are most often used because they offer smoothed out, continuous stress results. However, examination of element values provides very informative feedback on the quality of the results. If element values in two adjacent elements differ too much, as shown in figure 3-16, this indicates that the element size in this location is too large to properly model the stress gradient. By examining the element values, we can locate mesh deficiencies without running a convergence analysis. In this particular exercise with the L-bracket, the large differences in stresses, shown as element values, is the result of stress singularity which, of course, can not be helped by any amount mesh refinement.

To decide how much is "too much" of a difference requires some experience. As a general guideline, we may say that if the element values of stress in adjacent elements are apart by several colors on the color chart, then you should use a more refined mesh.

4: Analysis of a support bracket

Objectives

On completion of this exercise, you will be able to:

❑ Use shell elements for analysis of thin wall structures

❑ Perform a frequency analysis

Project description

We will analyze a stamped steel bracket, shown in figure 4-1, with the objective of finding displacements, stresses, and the first few modes of vibrations. This will require running both structural and modal analyses. We will use the SolidWorks model, called SUPPORT BRACKET.

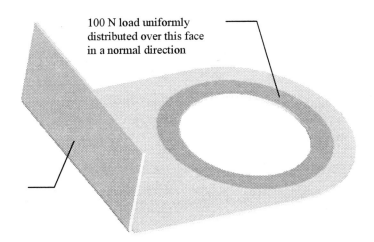

Figure 4-1: Support bracket

The thin bracket needs to be analyzed for displacements, stresses, and modes of vibrations.

Procedure

Before defining the study, consider that thin wall geometry would be difficult to mesh with solid elements. Generally it is recommended that two layers of second order tetrahedral elements be used across the thickness of a wall undergoing bending. In our case, this would require a large number of very small elements and the model would be slow to mesh and solve.

Therefore, instead of using solid elements, we will use shell elements to mesh the surface located mid-plane in the bracket thickness. The study definition with the meshing option selected for the mid-plane shell elements is shown in figure 4-2.

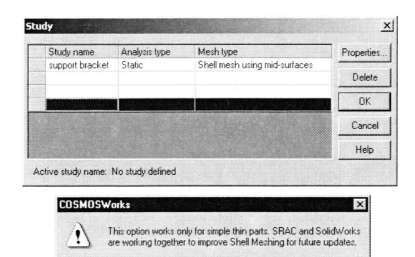

Figure 4-2: Study window and COSMOSWorks message

The study for the support brackets defines a shell mesh created mid-plane in the wall thickness. Also shown in figure 4-2 is a notification that this option works only for simple geometries.

Having decided to use shell elements for analysis, we assign material properties (Alloy Steel)—not to the *Solids* folder, which does not exist, but to the *Mid-surface Shell* folder as shown in figure 4-3.

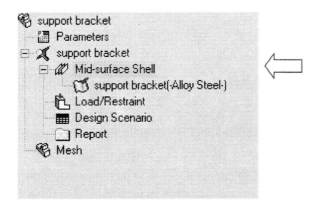

Figure 4-3: Assignment of material properties

Material properties are applied to shells if the design study has been defined as Shell mesh using mid-plane surfaces.

To apply loads, we select the face adjacent to the hole (figure 4-1) and apply 100 N normal force (figure 4-4).

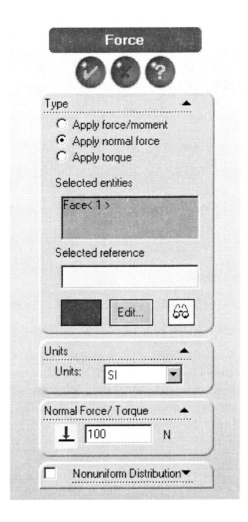

Figure 4-4: Force window

Support is applied to the vertical face, as shown in figure 4-1. Since shell elements have six degrees of freedom, there is now a difference between applying fixed supports or immovable supports. In our case, we need a fixed support, which eliminates all six degrees of freedom. Immovable support eliminates only translational degrees of freedom. The Restraint window is shown in figure 4-5.

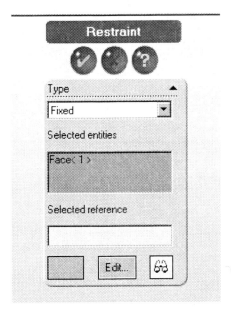

Figure 4-5: Restraint window

Note that fixed support rather than immovable support is selected.

Before creating the mesh, note that we do not explicitly define the shell thickness. COSMOSWorks assigns shell thickness automatically, based on the corresponding dimension of the solid CAD model, which in this case is 1.5 mm. In the Preferences window under the Mesh tab, we select **Automatic transition** The shell element mesh created with the default global element size (4.13 mm) and a mesh control element size (2.0 mm) applied to the round is shown in figure 4-6.

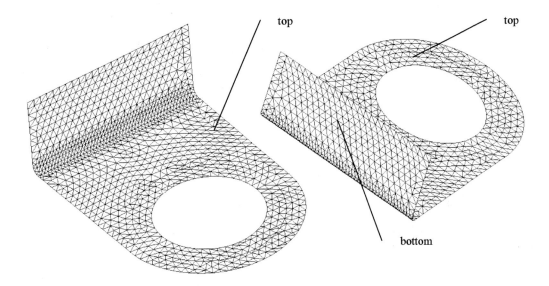

Figure 4-6: Shell element mesh

In the shell element mesh, elements have been placed mid surface between the faces that define the thin wall. Different colors distinguish between the top and bottom of the shell elements. The top is shown in gold and the bottom in gray. Those colors are shown here as different scales of gray.

The mid-plane where shell elements have been just created is shown as an imported feature in the SolidWorks Manager (figure 4-7). Also notice that shell element mesh uses different colors to distinguish between the top and bottom of the shell elements. You can modify the visual settings in the Preferences window under the Mesh tab (figure 4-8).

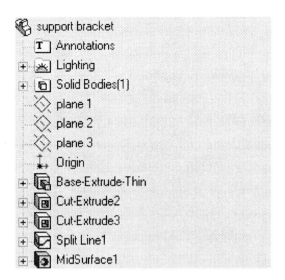

Figure 4-7: How the mesh is referenced in the SolidWorks Manager

The surface that has been meshed with shell elements is automatically created as an imported feature and can be seen in SolidWorks Manager.

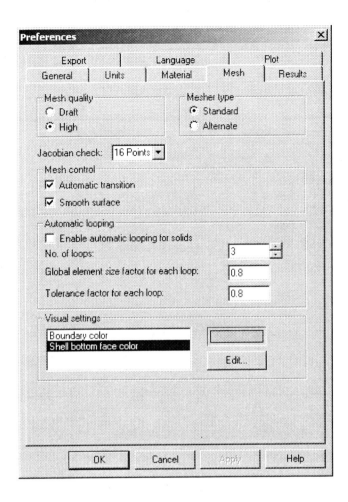

Figure 4-8: Preferences window

Color differentiates the top and bottom of shell elements. Specific colors can be assigned in the Visual settings area under the Mesh tab in the Preferences window.

Distinguishing between top and bottom of shell elements is very important when analyzing stress results. In the Stress plot window under the Display tab, you can select where the results are presented: at the top or at the bottom of shell elements (figure 4-9).

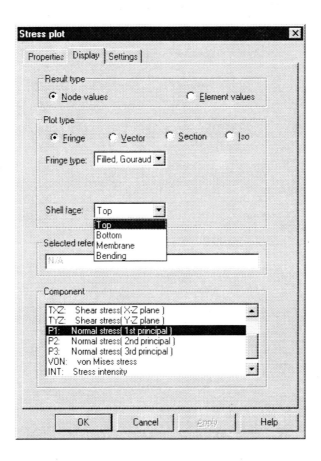

Figure 4-9: Selecting the location of stresses for shell elements

If shell elements are used, stress plot definition must include selection of the location where stresses are to be presented: at the top or at the bottom of shell elements. Also available are options to show membrane and bending components of stress.

Review the mesh colors to ensure that the shell elements are properly aligned. Try reversing shell element orientation by selecting the face where you want to reverse shell orientation. Then right-click the *Mesh* folder to display a pop-up menu, and select **Flip shell elements** (figure 4-10). Misaligned shell elements, that is, elements on the same side showing tops and bottoms facing the same way, lead to the creation of erroneous plots like one shown in figure 4-11 that shows a rectangular plate undergoing bending.

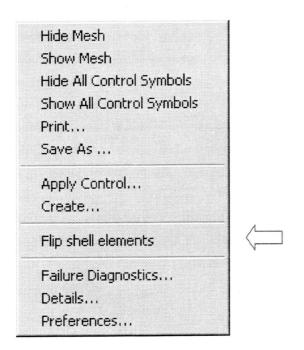

Figure 4-10: Pop-up menu for modifying shell element orientation

Reverse shell element orientation with this menu choice.

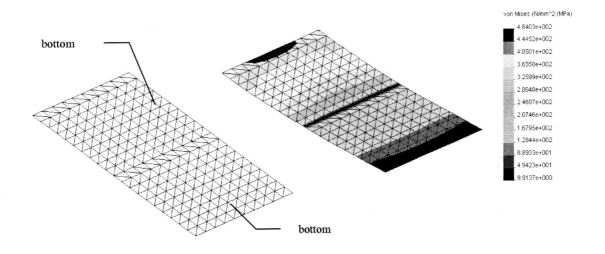

Figure 4-11: Erroneous shell element mesh and von Mises Stress plot resulting from faulty orientation of shell elements

The misaligned shell element mesh (left) and erroneous von Mises Stress plot are the result of shell element misalignment. This model is unrelated to our exercise.

The difference between stress results on top and bottom sides of the shell elements in our exercise model is illustrated in figures 4-12 and 4-13.

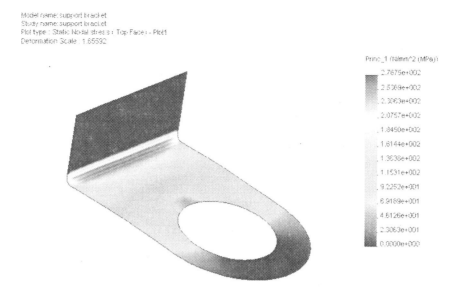

Figure 4-12: Maximum principle stress (P1) results for the top sides of the shell elements

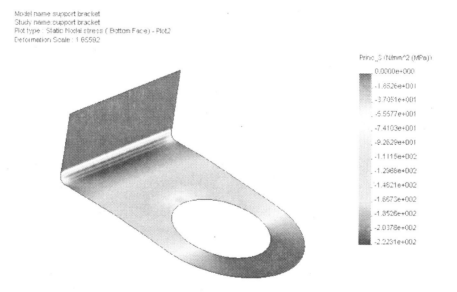

Figure 4-13: Minimum principle stress results for the bottom sides of the shell elements

Note that even though we are looking at the top sides of the shell elements (consult figure 4-6) due to the orientation of the model, the stress results are displayed for the bottom sides of the elements as if the shells were transparent.

Having obtained displacement and stress solutions, we have completed the structural analysis of the support bracket. We will now use the same model to calculate several natural frequencies for the same bracket. This calculation requires us to set up and run a frequency analysis (figure 4-14), often called a modal analysis.

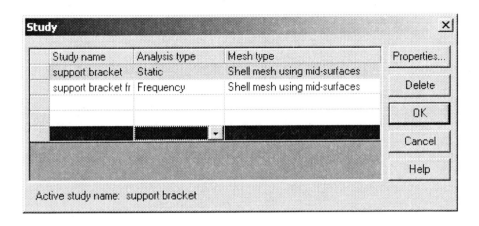

Figure 4-14: Study window showing two study definitions

Two studies are defined: the support bracket *study is a structural static analysis and the* support bracket fr *study is a frequency (or modal) analysis.*

You can copy the material definition for the frequency study from the static study by dragging and dropping the *Material Definition* icon into the corresponding item in the frequency study. The same can be done with supports. Notice that, although loads can be defined in a frequency study, they will have no effect. Frequency analysis calculates natural frequencies and modes of vibration, disregarding any effects of external loads, unless special options are activated (see chapter 18). Figure 4-15 shows the COSMOSWorks Manager window with both studies after solution.

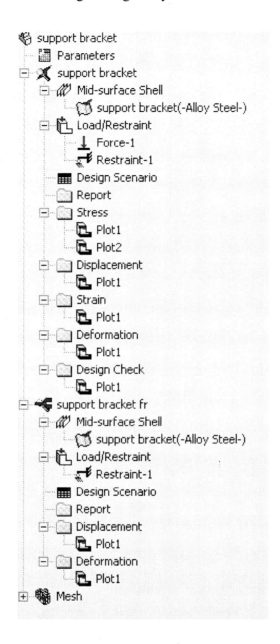

Figure 4-15: Two design studies defined and solved in the same model: the static (*support bracket*) study and the frequency (*support bracket fr*) study

No Stress *folder is created with a frequency analysis because a frequency analysis does not calculate stress results.*

Notice that, as shown in figure 4-15, the *Mesh* icon is not assigned to any particular design study. Therefore, the same mesh appears to be shared by different studies. Indeed, if we run the *support bracket fr* study to calculate frequencies, we will reuse the mesh that was originally created for the *support bracket* study. However, if we create a new mesh in the *support bracket fr* study, the new mesh will not overwrite the first mesh. Depending on which design study was last accessed, the **Show Mesh** command will display the appropriate mesh.

We can prove this fact by deleting the mesh control before creating a mesh for the *support bracket fr* study and creating the new mesh without bias in the round. Frequency analysis does not analyze stresses; therefore, the mesh bias that was necessary to produce accurate stress results is permissible, but not required, in the frequency analysis. To become familiar with the **Show Mesh** feature, display the two different meshes by alternately activating the *structural* and *frequency* studies, and then selecting the **Show Mesh** command for each study.

Before we actually run the Frequency analysis, let's review the properties of a frequency study. The Frequency window offers the choice of how many frequencies and associated modes of vibration will be calculated (figure 4-16). We accept the default number of five frequencies.

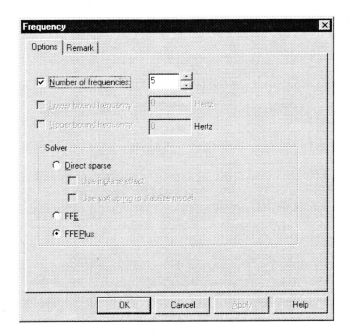

Figure 4-16: Frequency window

The Options tab in the Frequency window allows you to select properties of the frequency study. Calculation of the first five natural frequencies is the default request.

When a frequency analysis is run, two *Results* folders are automatically created: a *Displacement* folder and a *Deformation* folder. Each folder holds one plot, which, by default, shows the results for the first frequency (figure 4-17). Figure 4-18 shows displacement results related to this first frequency, or *Mode 1*.

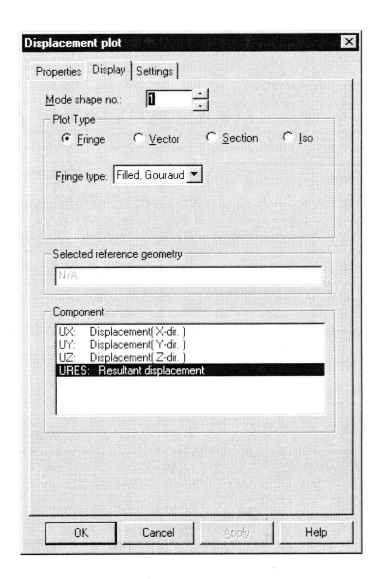

Figure 4-17: Displacement plot window

The definition of the displacement plot in the Frequency analysis requires you to specify the mode to be displayed. Here we select the first mode or Mode 1.

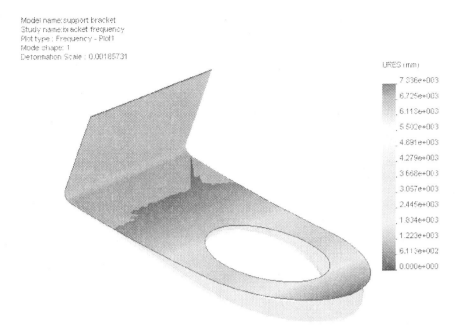

Figure 4-18: Displacements associated with *Mode 1*

Note that the undeformed shape is superimposed on the deformed shape.

Notice that the displacement plot in figure 4-18 shows extremely high displacement magnitudes. These unrealistic displacement magnitudes require an explanation. Frequency analysis does not provide any quantitative information on displacements. All that can be learned from the plot in figure 4-18 are relative displacements. In fact, displacement plots are rarely used to analyze the results of frequency studies. Displacement results are purely qualitative and can be used for comparison within the same mode of vibrations. Even relative comparison of displacements between different modes makes no sense.

More informative, and less confusing, is the deformation plot. The deformation plot shows the shape of deformation (with no fringes) associated with the given mode of vibration as well as the associated frequency of vibration (figure 4-19).

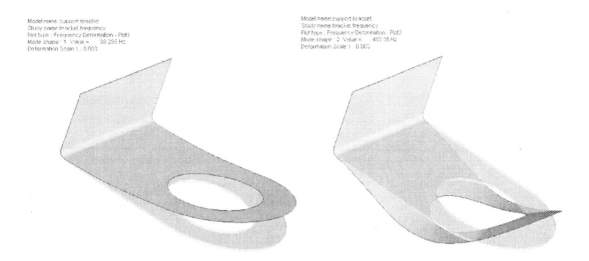

Figure 4-19: Deformation plot showing the shape of deformation and frequency of vibration for a given mode

The image on the left illustrates the first mode of vibration with a frequency of 88 Hz, and the image on the right shows the third mode with a frequency 403 Hz. You can display the undeformed shape superimposed on the deformed shape in the Deformed Shape Plot definition window (right- click the Plot *icon and select* **Edit definition…***).*

By far the best way to analyze the results of a Frequency analysis is by examining the Animated Deformation plots. Try creating Deformation plots for all five calculated frequencies and review the animated plots. To animate any plot, right-click a *Plot* icon to display an associated pop-up menu, and then select **Animate**.

5: Analysis of a Link

Objectives

On completion of this exercise, you will be able to:

- Prepare a SolidWorks CAD model for analysis with COSMOSWorks
- Define symmetry boundary conditions
- Modify restraints
- Prevent rigid body motions in the model

Project description

In this exercise, we need to calculate the displacements and stresses of a steel link shown in figure 5-1. The link is pin-supported in two end holes and is loaded with 100,000 N applied to the central hole. The other two holes are not loaded. For this exercise, we will use the SolidWorks model called LINK.

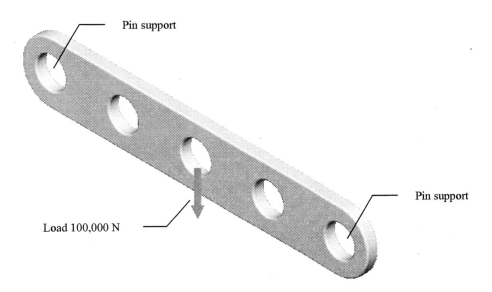

Figure 5-1: CAD model of the link

Note that the supporting pins and the pin, through which load is applied, are not shown. All rounds need to be suppressed to simplify meshing.

Procedure

One way to conduct this analysis would be to model both the link and all three pins, and then conduct an assembly analysis. However, we are not interested in the contact stresses that will develop between pins and the link. Our focus is on the deflections and bending stresses that will develop in the link; therefore, the analysis can be simplified by *not* modeling the pins. Instead, we can simulate the pins by proper definition of restraints and load. Further, notice that the link geometry, supports, and load are symmetrical. We can take advantage of that symmetry to analyze only half of the model by replacing the other half with symmetry boundary conditions. Make sure that cut (the last feature in SolidWorks Feature Manager window) is not suppressed.

Before proceeding, we need to suppress all small rounds in the SolidWorks model. Those rounds have negligible structural effect and would unnecessarily complicate the finite element mesh. Removing geometry details that are deemed unnecessary for analysis is called defeaturing.

Finally, we note a split face in the middle hole that defines the area where load will be applied. Geometry in FEA-ready form is shown in figure 5-2. Figure 5-2 also explains what restrains need to be applied.

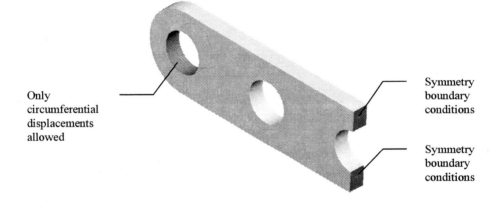

Figure 5-2: Geometry ready for COSMOSWorks

Note that the model has been cut in half and small rounds have been suppressed. Highlighted in green are:

❑ Hole where pin support is simulated

❑ Two faces in the plane of symmetry where symmetry boundary conditions are required

We can now move to COSMOSWorks, where we can define a study (static analysis, solid element mesh) and apply material properties of alloy steel. Now, the model is ready for the definition of supports, the highlight of this exercise.

To select a cylindrical face, as shown in figure 5-2, right-click the *Load/Restraint* icon to display a related pop-up menu, and then select **Restraint**. This action opens the Restraint window shown in figure 5-3.

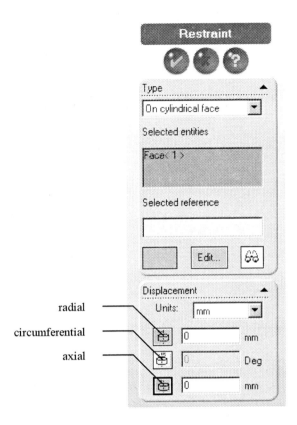

Figure 5-3: Restraint window

The restraint definition window specifies the restraint type as **On cylindrical face** because **Cylindrical Face** was selected prior to opening the Restraint window. Notice that restraint directions are now associated with the cylindrical face directions (radial, circumferential, and axial), rather than with global directions (x, y, z).

To simulate the pin support that allows the link to rotate about the pin axis, displacement in the radial direction needs to be restrained and displacement in the circumferential direction needs to be allowed. Moreover, displacement the in axial direction needs to be restrained in order to avoid rigid body motions of the entire link along this direction. Notice that while we must restrict rigid body motion of the link in the direction defined by the pin axis, we can do this by restraining any point of the model. We simply find it convenient to apply the axial restraints to this cylindrical face.

As shown in figure 5-2, the analyzed geometry consists of one half of the link. To simulate the entire link, we apply symmetry boundary conditions to two faces located in the plane of symmetry. The symmetry boundary conditions require that these two faces remain in the plane of symmetry while the link experiences deformation, meaning that in-plane displacements are allowed, but out-of-plane displacements must be suppressed. The easiest way to define symmetry boundary conditions is to use **On flat face** restraints to restrain displacement in the direction normal to the flat face. The definition of the symmetry boundary conditions is illustrated in figure 5-4.

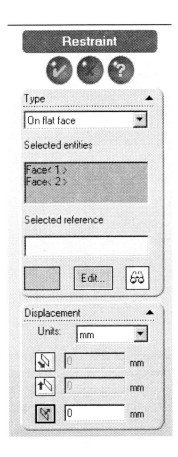

Figure 5-4: Definition of symmetry boundary conditions

Recall from figure 5-1 that the link is loaded with 100,000 N. Since we are modeling half of link, we need to apply only 50,000 N load to a portion of the cylindrical face, as shown in figure 5-5. The size of area where load is applied is arbitrary, but close to what we would expect to be the contact area between the pin and the link.

When defining the load (in reference to figure 5-5), it is important to remember to first select a reference plane from the SolidWorks Manager (here, we use the **Top** plane for reference), then return to COSMOSWorks Manager, and right-click the *Load/Restraint* folder. This way, the selected (Top) reference plane will be used as the reference geometry for defining the direction of force. Force direction should be defined as normal to the reference plane. In order not to lose the selected reference plane, press and hold the *Crtl-* key until the Force window opens. Once the Force window opens, you can select the face where force should be applied. Alternately, you may select the reference place and the face (also using *Crtl-* key), and then open the Force window.

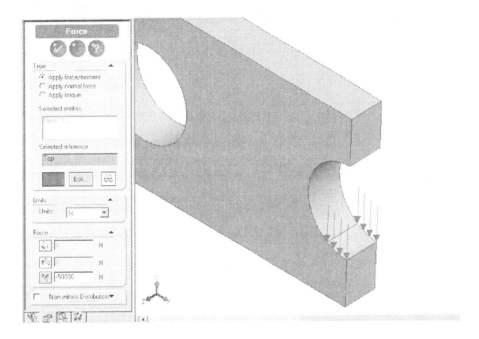

Figure 5-5: 50,000N force applied to the central hole

*The **Top** reference plane is used to determine the load direction. Notice that load is distributed uniformly as we do not attempt to simulate contact stress problems.*

The last task of model preparation is meshing. Right-click the *Mesh* icon to display the related pop-up menu, and then select **Create....** Verify that the mesh preferences are set on high quality, meaning that second order elements will be created. Next, mesh the geometry using the default element size suggested by the automesher. For more information on the mesh, you may wish to review Mesh details (figure 5-6).

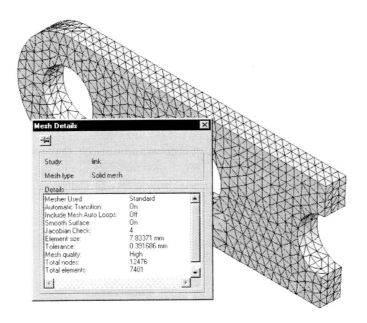

Figure 5-6: Mesh Details window shown over the meshed model

After solving the model, we first need to check if the applied boundary conditions work properly, meaning whether the link can rotate around the pin and that it behaves as if the whole link, and not only half of it, were present. This verification is best done by examining the animated displacements, preferably with both undeformed and deformed shapes visible (figure 5-7).

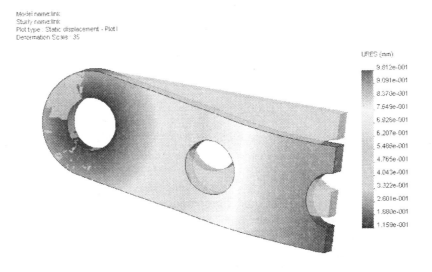

Figure 5-7: Comparison of the deformed and undeformed shapes

Comparison of the deformed and undeformed shapes verifies the correctness of the restraints definition.

To conclude this exercise, we will review the stress results. Figure 5-8 shows von Mises stress results. Please examine the different stress components, including the maximum principal stresses, minimum principal stresses, etc.

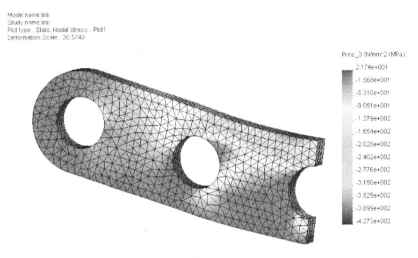

Figure 5-8: Sample of stress results

In this sample, the minimum principal stress (P3) results are shown with mesh superimposed on the fringe plot.

You may also wish to repeat this exercise using the full model. To do this, open the SolidWorks Manager, suppress the cut, and return to COSMOSWorks. You will be prompted to acknowledge the change in the model and the, now invalid, restraints that were earlier used to define the symmetry boundary conditions (figure 5-9). You will need to add load to the now complete model, define support for the other pin, and then proceed with meshing, running the solution, and reviewing results analysis.

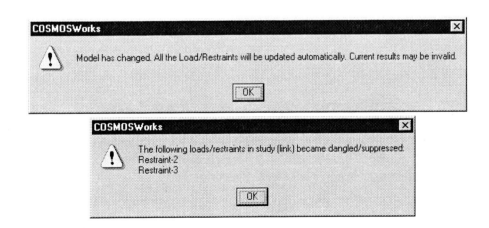

Figure 5-9: COSMOSWorks notifies the user of changes in model geometry

Any change in model geometry invalidates FEA results. Changes may also require modification of restraints and loads definition.

Before finishing the analysis of LINK we should notice that link supported by two hinges, as modeled in this exercise, corresponds to configuration shown in figure 5-10 where one of the hinges is floating. Since linear analysis does not account for changes in model stiffness during the deformation process, linear analysis is unable to model membrane stresses that would have developed if both hinges were in fixed position. Non linear geometry analysis would be required to model support if both hinges were in fixed position. See again chapter 1 for a brief review of limitations of linear analysis.

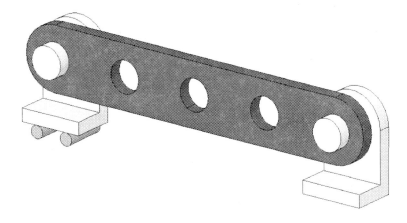

Figure 5-10: Our model corresponds to the situation where one hinge is floating as symbolically shown here by rollers under the left hinge.

Non linear geometry analysis would be required to model supports if both hinges were in fixed positions.

6: Analysis of a tuning fork

Objectives

On completion of this exercise, you will be able to:

❑ Perform a frequency analysis with and without supports

❑ Discuss the role of supports in a frequency analysis

Project description

As you have noticed, all exercises in this book go beyond purely software-specific instructions. We also cover the FEA background necessary to understand the meaning of the undertaken exercises. The present exercise has very low "software-specific" content. In this chapter, we analyze a tuning fork in order to gain more insight into a frequency analysis, also called a modal analysis.

As we know, each structure has preferred frequencies of vibration, called resonant or modal frequencies. When excited with a resonant frequency, a structure will vibrate in a certain shape, called mode of vibration. The only factor controlling vibration amplitude is damping. While every real life structure has an infinite number of modal frequencies and associated modes of vibration, only a few of the lowest modes are important in their response to dynamic loading. A frequency analysis calculates those resonant frequencies and their associated modes of vibration.

Open the SolidWorks file called TUNING FORK, and review its geometry (figure 6-1).

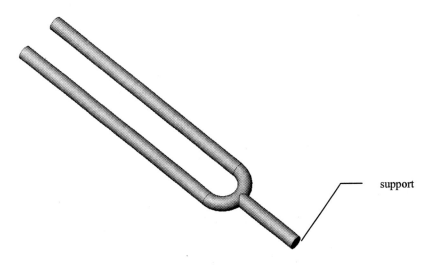

Figure 6-1: Tuning fork geometry

Support is applied to the round end-face.

Procedure

A quick inspection of the CAD geometry reveals a sharp re-entrant edge. This condition renders the geometry unsuitable for stress analysis, but acceptable for frequency analysis. As any musician will tell us, the tuning fork produces a low *A* sound, which has a frequency of 440 Hz. Let's run a frequency analysis and verify that the fundamental frequency is indeed 440 Hz. To do this, we will create a study with supports as a frequency analysis with solid elements, and then request that four natural frequencies in the tuning fork be calculated (figure 6-2).

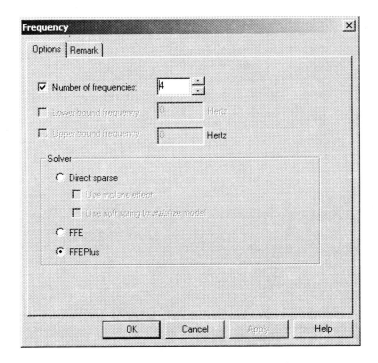

Figure 6-2: Frequency window

The Options tab allows you to select the number of frequencies to be calculated in the frequency study, as well as the solver. Notice that, as always, different solvers are available. We request that four frequencies be calculated.

The frequency study will calculate the first four natural frequencies of the tuning fork using the default solver, FFEPlus, which is also the fastest solver.

Next, we assign the properties of alloy steel to the only solid present in the study and apply support by defining restraints to the end face, as shown in figure 6-1. This support approximates the situation when the tuning fork is held in two fingers. Either Immovable or Fixed restraints can be applied (solid elements, as we now know, do not distinguish between those two types of restraints).

Finally, we mesh the model with a default element of 1.44 mm and review the meshing details.

The meshed model is shown in figure 6-3. The COSMOSWorks automesher is tuned to the requirements of a stress analysis. A frequency analysis is less demanding on the mesh. Generally a less refined mesh is acceptable for frequency analysis as compared to that required for a stress analysis of the same model. Nevertheless, since the model size is small anyway, we will use the mesh that we created with the default element size.

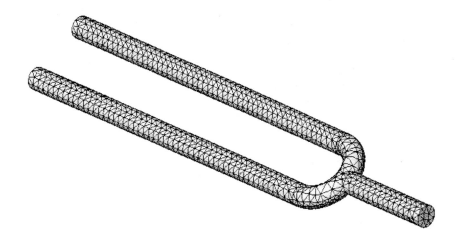

Figure 6-3: Meshed model of the tuning fork ready for frequency analysis

After the solution is complete, COSMOSWorks creates two *Results* folders: *Displacement* and *Deformation* and places one plot in each folder, showing the displacements and deformations corresponding to the first mode of vibrations. If desired, you can define more results plots by right-clicking the *Displacement* or *Deformation* folders, to open the associated pop-up menu, from where you can define more results plots. We will add three more plots to the *Deformation* folder, corresponding to *Mode 2*, *Mode 3*, and *Mode 4*. You can define or modify plot properties in the Deformed Shape Plot window, which opens during plot definition (figure 6-4) or by editing a plot (right-click a *Plot* icon).

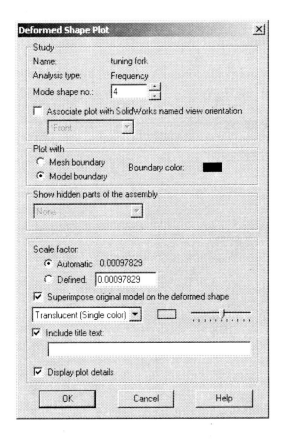

Figure 6-4: Deformed Shape Plot window

This window allows you to display deformations with the original shape of the model superimposed on the deformed shape.

Having defined four plots, we can give them more meaningful names using the already-discussed Click-inside technique. The renamed deformation plots (figure 6-5) now correspond to their associated mode number. Plots can be displayed by either:

❑ Double-clicking the appropriate *Plot* icon, or

❑ Right-clicking a *Plot* icon to display the associated pop-up menu, and then clicking **Show**.

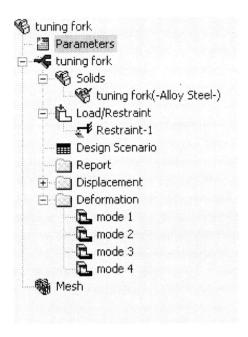

Figure 6-5: *Deformation* folder with the four plots renamed to correspond with the number of the mode of vibration they represent

The deformation plot in the modal analysis shows the mode of vibration associated with a given natural frequency. All four modes of vibration are presented in figure 6-6.

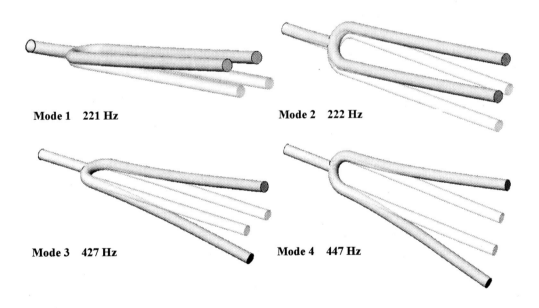

Figure 6-6: First four modes of vibration and their associated frequencies

We notice that the lower *A* frequency of 440 Hz, which we were expecting to be the first mode, is actually the fourth mode. Before explaining the reasons for that, let's run the frequency analysis once more, this time without any restraints. We need to define a new frequency study, which we'll call *tuning fork no supports*. We only need to define the material properties and can do this by either:

❑ Selecting the appropriate library file, or

❑ Copying them from the study that we named tuning fork.

In the Properties window of the *tuning fork no supports* study, we again specify that four modes be calculated. After the solution is completed, we right-click the automatically created *plot1* that has been placed in the *Deformation* folder and select **Edit Definition…**. Much to our surprise, we find that the highest mode that can be specified is mode 10 (figure 6-7) even though we asked that only four modes be calculated.

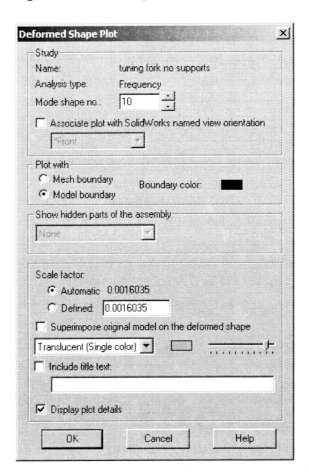

Figure 6-7: Deformed Shape Plot window

Note that the Mode shape number in the Deformed Shape Plot window is the 10th node of vibration, even though only four modes were requested in the Properties window of the tuning fork no supports *study.*

To find out why ten modes, not four, were calculated, we need to define plots for all ten modes. We then notice that the first six modes have the associated frequency 0 Hz. Why? The first six modes of vibration correspond to rigid body modes. Because the tuning fork is not supported, it has six degrees of freedom as a rigid body: three translations and three rotations.

COSMOSWorks detects those rigid body modes and assigns them zero frequency (0 Hz). The first elastic mode of vibration, meaning the first mode requiring that the tuning fork experience elastic deformation, is *Mode 7*. *Mode 7* has a frequency of 447 Hz, close to what we were expecting to find as the fundamental mode of vibration for the tuning fork.

Why did the frequency analysis with support not produce the first mode with a frequency close to 440 Hz, the frequency with which the tuning fork is supposed to vibrate? If we closely examine the first three modes of vibration of the supported tuning fork, we notice that all three need support in order to exist. As the analysis with no supports proved, without support, the tuning fork will not vibrate in those three modes. Because support is needed to sustain those modes, vibrations in the first three modes are quickly damped by support and the tuning fork, with or without support, eventually vibrates the way it was designed to. Indeed, notice that *Mode 4*, (calculated in the analysis with supports) and *Mode 7* (calculated in the analysis without supports) are identical.

NOTES:

7: Analysis of crossing pipes

Objectives

On completion of this exercise, you will be able to:

- Perform a simple thermal analysis
- Draw analogies between structural and thermal analysis
- Explain temperature distribution due to prescribed temperatures

Project description

So far, we have practiced static analyses and frequency analyses. Both deal with structural properties of the examined model. Static analysis provides results in the form of displacements, strains, and stresses. Frequency analysis provides results in the form of natural frequencies and the associated modes of vibration. We will now examine a thermal analysis. Numerous analogies exist between thermal and structural analyses. The most direct analogies are summarized in figure 7-1.

Structural Analysis	**Thermal Analysis**
Displacement [m]	Temperature [$°C$]
Strain [1]	Temperature gradient [$°C/m$]
Stress [N/m^2]	Heat flux [W/m^2]
Load [N] [N/m] [N/m^2] [N/m^3]	Heat source [W] [W/m] [W/m^2] [W/m^3]
Prescribed displacement	Prescribed temperature

Figure 7-1: Analogies between structural and thermal analysis

Note that SI units are used.

In this exercise, we will perform a simple thermal analysis of two pipes crossing, using the SolidWorks file named CROSSING PIPES.

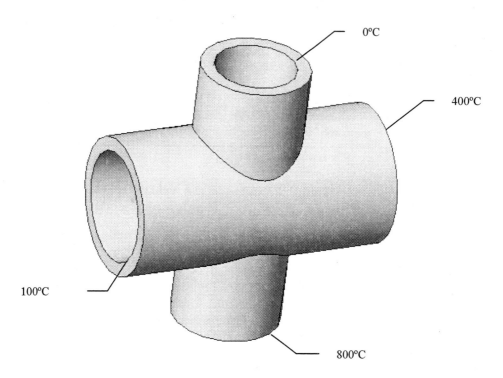

Figure 7-2: CAD model of two crossing pipes

Also shown are the prescribed temperatures, which we will apply as boundary conditions in the thermal analysis.

Procedure

Temperatures to be applied to the model are shown in figure 7-2. As indicated in figure 7-1, prescribed temperatures are analogical to prescribed displacements in structural analyses. There are no other thermal conditions in the model; no heat sources and no convection coefficients. Our objective is to determine what temperature field will establish itself once the prescribed temperatures are applied. Note that heat will flow in the model due to the temperature gradient, but, since no convection coefficients are defined on any faces, no heat will escape the model. In thermodynamics jargon, we say that the model has *adiabatic* walls. Also note that the model geometry is simplified (defeatured) and has sharp re-entrant corners. The defeatured model is acceptable only if all we want to determine is the temperature distribution. Sharp re-entrant corners are *not* acceptable for the analysis of heat flux for the same reasons that they are not acceptable for the analysis of stresses in a structural analysis.

The first step is, as always, a definition of the study. Call this study *crossing pipes* and define it as shown in figure 7-3. Next, apply the material properties of **Grey Cast Iron** from the COSMOSWorks material library.

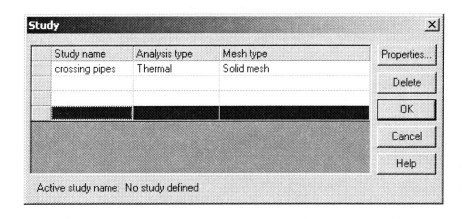

Figure 7-3: Definition of *crossing pipes* study

The study definition for crossing pipes *is a thermal analysis type using a solid mesh.*

To define the temperature, right-click the *Loads/Restraints* folder to open a pop-up menu specific for thermal analysis (figure 7-4), and select **Temperature….**

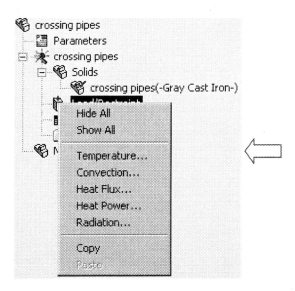

Figure 7-4: Pop-up menu related to a thermal analysis

The pop-up menu lists the thermal boundary conditions available in a thermal analysis.

Figure 7-5 shows the Temperature window for the face where a temperature of 100°C is applied. Since each of four faces has a different prescribed temperature, we need to assign temperatures in four separate steps. In the Temperature window, we enter the prescribed temperature definition.

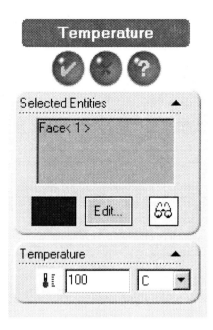

Figure 7-5: Temperature window

Notice that the units are in degrees Celsius.

The next step is meshing the model. Use the default element size and create a mesh, as shown in figure 7-6.

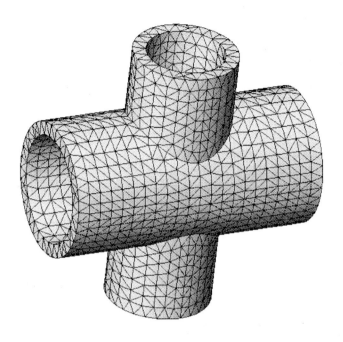

Figure 7-6: Finite element mesh of crossing pipes model created with the default element size

After solving the model, we notice that only one *Solution* folder has been created (figure 7-7). The folder, called *Thermal*, has one plot in it. By default, this automatically created plot shows the temperature distribution (figure 7-8).

Due to the deficiencies of the model geometry (sharp re-entrant corners), we must limit the analysis of results to examining temperature distribution only.

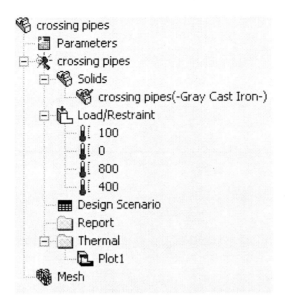

Figure 7-7: COSMOSWorks Manager window and the single *Thermal* folder

The COSMOSWorks Manager window shows only one *results* folder, called *Thermal*. Notice that we have renamed the prescribed *temperature* icons to give them more descriptive names.

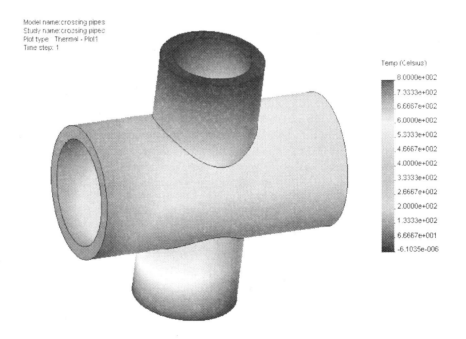

Figure 7-8: Temperature distribution in the crossing pipes model

8: Analysis of a radiator assembly

Objectives

On completion of this exercise, you will be able to:

- Perform an analysis of assemblies
- Perform a thermal analysis with heat power and convection coefficients
- Create section views in results plots

Project description

In this exercise, we continue with thermal analysis; however, this time, we will analyze an assembly rather than a single part, as we have done until now. Let's open the SolidWorks assemblies file named RADIATOR ASSEMBLY and examine it (figure 8-1).

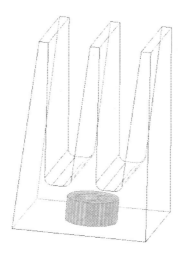

Figure 8-1: CAD model of a radiator assembly

The CAD model of a radiator assembly consists of two components: aluminum radiator, shown here as transparent, and a ceramic insert representing a microchip.

The ceramic insert generates a heat power of 50 W. The heat is dissipated by the aluminum radiator. The ambient temperature is 27°C. Heat is dissipated to environment by convection through all surfaces, except the base, which is isolated. The convection coefficient, also called the film coefficient, is 25 W/m^2/°C. This particular value means that if the difference of temperature between the face of the radiator and the surrounding air is 1°C, then each square meter of the surface dissipates 25 W of heat.

Our objective is to determine the temperature of the outside faces of the aluminum radiator as well as temperature on the border between the radiator and the ceramic insert.

Procedure

Let's now move from SolidWorks to COSMOSWorks and create a study called *heat sink*. Before proceeding further, we need to investigate a new icon, called *Contact/Gaps*, which is found in the COSMOSWorks Manager window (figure 8-2).

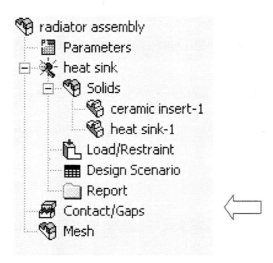

Figure 8-2: *Contact/Gaps* icon in the COSMOSWorks Manager

When an assembly is analyzed, the COSMOSWorks Manager window always includes a Contact/Gaps *icon.*

Right-click the *Contact/Gaps* icon to open the pop-up menu, shown in figure 8-3.

Figure 8-3: Pop-up menu for defining and editing contact/Gaps conditions

As figure 8-3 shows, the default setting for Contact/Gaps conditions is **Touching Faces: Bonded**, meaning that assembly components are merged. This setting is exactly what we need; therefore, nothing needs to be modified in the Contact/Gaps settings. We will investigate Contact/Gaps conditions more closely in further examples.

Analysis of an assembly allows assigning different material properties to each assembly component. Notice that the *Solids* folder, visible in figure 8-2, contains two icons corresponding to two assembly components. By right-clicking each one, we open an associated pop-up menu and assign **Ceramic Porcelain** material to the ceramic insert assembly component and **Aluminum Alloy 1060** to the radiator assembly component.

Next, we need to specify the heat power generated in the porcelain insert. To do this, select the *ceramic insert* icon from the *Solids* folder. Press and hold the *Ctrl*-key and right-click the *Load/Restraint* folder to open the pop-up menu, shown in figure 8-4. From the pop-up menu, select **Heat Power…** to open the Heat Power window, as shown in figure 8-5.

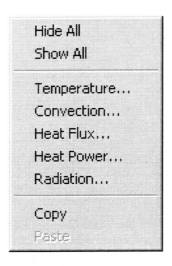

Figure 8-4: Pop-up menu associated with the *Load/Restraint* folder

Right-clicking the Load/Restraint *folder opens this pop-up menu, which lists the available loads and restraint choices in a thermal analysis.*

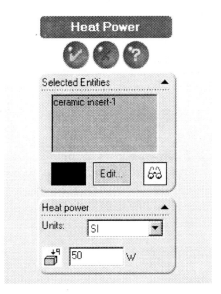

Figure 8-5: Heat Power window

*Notice that **ceramic insert-1** appears as the Selected Entity.*

So far we have assigned material properties (different for each component) and a heat source, but in order for heat to flow, we must establish a mechanism for heat to escape from the model. We can accomplish this by defining convection coefficients on the external faces of the radiator.

Let's then select all faces (except for the base) of the radiator. Next, right-click the *Load/Restraint* folder to open a pop-up the menu already shown in figure 8-4. From the pop-up menu, select **Convection...** to open the Convection window, shown in figure 8-6.

Engineering Analysis with COSMOSWorks

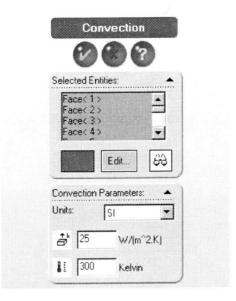

Figure 8-6: Convection window

This window allows you to specify convection parameters: units, convection coefficient, and ambient temperature.

We specify a convection coefficient of 25 W/m^2/K for all selected faces and define the ambient temperature as 300K, which is equivalent to 27°C, the specified ambient temperature.

The last step before solving is creating the mesh. We can use the default element size of 4.8 mm to create the mesh shown in figure 8-7.

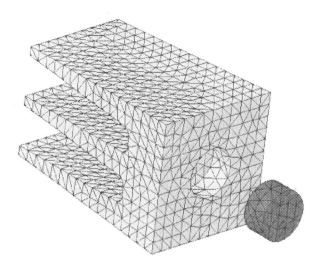

Figure 8-7: Exploded view of the meshed assembly

The exploded view is created using the SolidWorks Configuration Manager in the same way as any other exploded view in SolidWorks.

Once the solution is ready, we will define two Results plots: thermal and heat flux. The choices we made in the definitions of those plots are shown in figures 8-8 and 8-9.

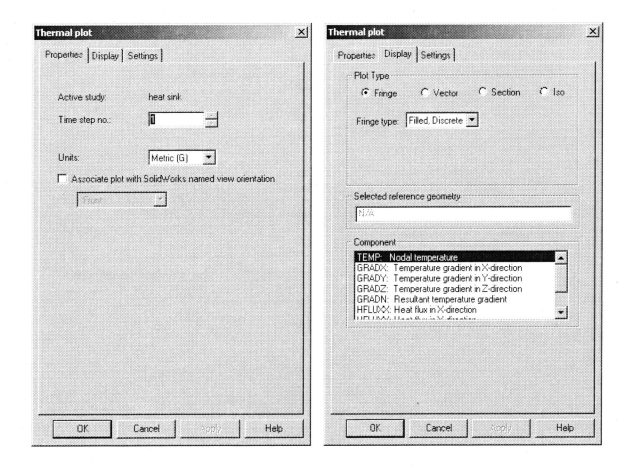

Figure 8-8: Definition of the thermal plot: Properties tab and Display tab

Note that we use metric units in order to review temperature distribution in °C rather than in K.

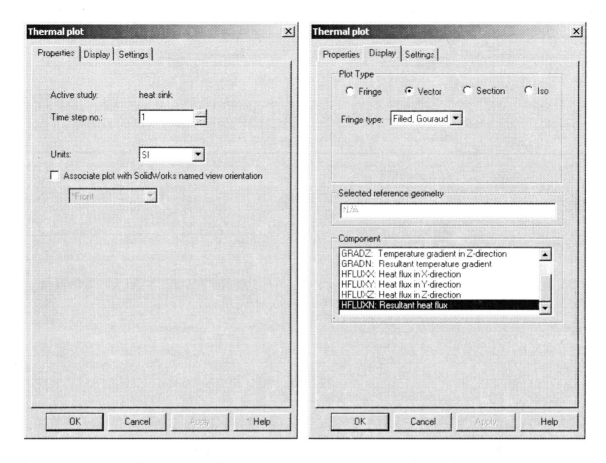

Figure 8-9: Definition of the heat flux plot: Properties tab and Display tab

Note that here we use SI units. We selected the Vector Plot type.

The resulting thermal and heat flux plots are shown in figures 8-10 and 8-11.

Figure 8-10: Temperature distribution in the assembly

The temperature of the ceramic insert core reaches 327°C while the outside faces of the radiator are at 105°C.

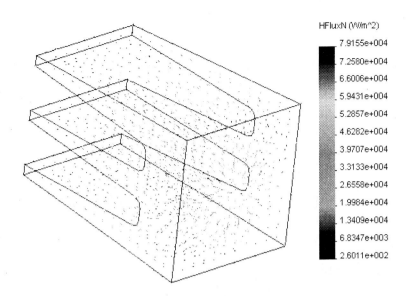

Figure 8-11: Heat flux in the assembly

The highest heat flux, 80,000W/m^2, is found on the interface between the ceramic insert and the radiator.

As figure 8-11 indicates, the location of the highest heat flux is inside the model. We will present two techniques that allow the display of those results more clearly. One way to display results on the interface between the insert and the radiator is to hide one of the assembly components. We can do that in the SolidWorks Manager. After this adjustment, the heat flux plot will look like that in figure 8-12.

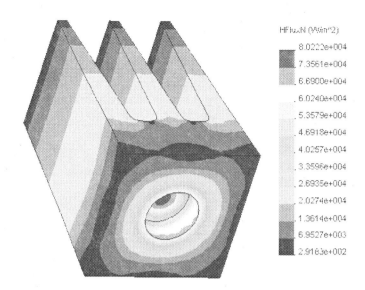

Figure 8-12: Heat flux results after hiding one of the assembly components

Hiding the ceramic insert reveals the heat flux pattern inside the assembly. "Unevenness" of the fringes along the edge results is due to the rather coarse mesh. A more refined mesh is necessary if we were interested in detailed results for this location.

Another way to access results inside the model is to specify a section view (figure 8-13).

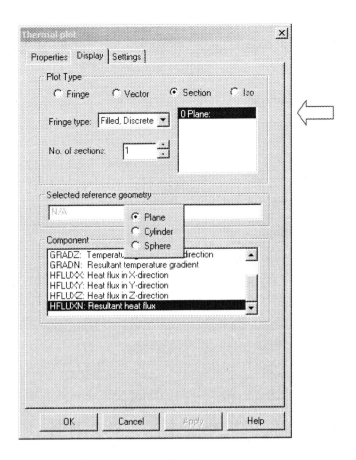

Figure 8-13: Thermal Plot window, Display tab: Specifying a section view

The plot can be specified as a Section plot by selecting the appropriate button in the Plot Type area. Then you need to specify the cutting plane. To do this, double-click the Plot Type area indicated by the arrow. This action opens a pop-up menu, which you can use to specify the cutting plane. Note that you can also define the number of cutting planes to be used.

We have defined the plot type as **Section**, but we still need to define the location of the cutting plane. To do this, right-click the *Plot* icon to open a pop-up menu (figure 8-14), and select **Clipping....** This action opens the Section Clipping window (figure 8-14).

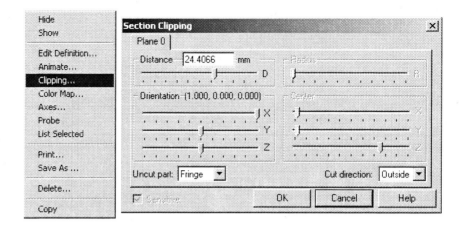

Figure 8-14: Pop-up menu (left) used to open the Section Clipping window (right)

*Select **Clipping...** from the pop up menu (left) to open the Section Clipping window.*

The position of the clipping plane as defined in the Section Clipping window relates to the global coordinate system. Orientation is defined by the position of the vector normal to the clipping plane. Both the Distance and Plane Orientation need to be specified for the desired results. These specifications depend on the model location in the global coordinate system.

The Orientation coordinates shown in figure 8-14 produce the Section plot shown in figure 8-15.

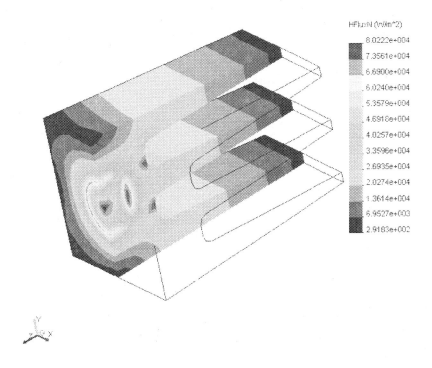

Figure 8-15: Section plot revealing high heat flux inside the model

Before you continue, please experiment with different types of cutting planes and different plane orientations.

NOTES:

9: Analysis of a hanger assembly

Objectives

On completion of this exercise, you will be able to:

- Perform an assembly analysis
- Apply global and local Contact/Gaps conditions

Project description

In this exercise we will investigate the structural analysis of assemblies, but first we need to review different options available for defining the interactions between assembly components.

Let's look more closely at the pop-up menu that opens when you right-click the *Contact/Gaps* icon, shown in figure 8-3 and repeated in figure 9-1 for easy reference.

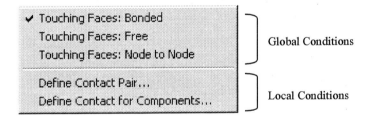

Figure 9-1: Pop-up menu associated with the *Contact/Gaps* icon

The pop-up menu, opened by right-clicking the Contact/Gaps *icon, distinguishes between global and local Contact/Gaps conditions.*

Notice that the default choice, as shown in figure 9-1, is that touching faces are bonded.

Notice a dividing line in the pop-up menu separating global Contact/Gaps conditions: **Touching Faces: Bonded**, **Touching Faces: Free**, and **Touching Faces: Node to Node** from local Contact/Gaps conditions.

- Global condition—affects all faces in an assembly
- Local condition—affects two specified faces or components in an assembly

A description of the available options follows:

Contact/Gaps Conditions	Option	Description
Global	Touching Faces: Bonded	The default choice, select this option when all touching faces are merged and assembly behaves as one part. If touching faces are left as **Bonded,** the only difference between part and assembly is that in assembly we can assign different material properties to individual components, while in part the entire model must have the same material properties.
	Touching Faces: Free	Select this option when the assembly is a series of unattached components with no structural connection between them
	Touching Faces: Node to Node	Select this option when touching faces can come apart but cannot penetrate each other
Local	Define Contact Pair	Select this option for a pair of surfaces in contact to define Contact/Gaps conditions individually for each face
	Define Contact for Components	Select this option for two components in contact to define Contact/Gaps conditions individually for each component

Global Contact/Gaps conditions can be overridden by component and/or local conditions. For example, using global Contact/Gaps conditions, we can request that all faces be bonded. Next, we can locally override this condition and define local conditions for one (or more) pair(s) as **Touching Faces: Node to Node**. The hierarchy of global, component, and local Contact/Gaps conditions is shown in figure 9-2.

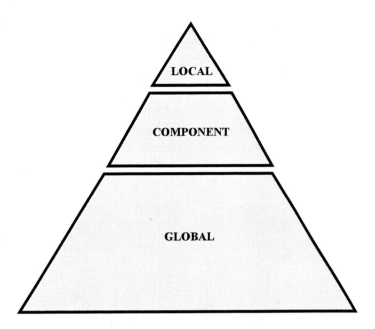

Figure 9-2: Hierarchy of Contact/Gaps conditions

In the hierarchy, local conditions override component and global conditions. Component conditions override global conditions.

Local Contact/Gaps conditions can be specified as Bonded, Free, Node to Node, Surface, and Shrink Fit. We'll now discuss the important distinction between Node to Node and Surface conditions.

Node to Node conditions, that can be specified both globally and locally, can be applied to faces that are identical in shape even if the faces are not the same size. Node to Node conditions can be specified between two:

- Flat faces
- Cylindrical faces of the same radius
- Spherical faces of the same radius

With these options, the mesh on both faces in the area where they touch each other is created identically and there is be node to node correspondence on both touching surfaces; hence the name, Node to Node.

Surface contact, which is specified between two touching surfaces of different shapes, can be only specified as a local condition. The mesh on both surfaces is not identical and there is no node to node correspondence.

The difference between Node to Node and Surface conditions is illustrated in figure 9-3.

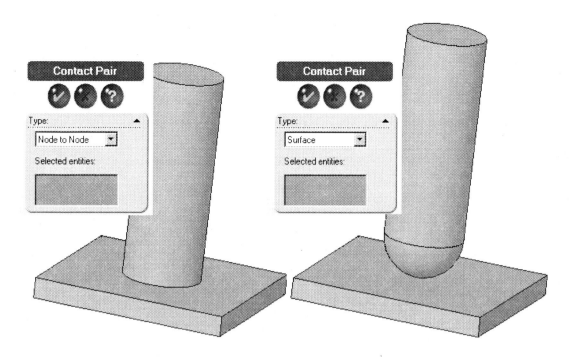

Figure 9-3: Contacts between a flat end punch / spherical punch and a flat plate

The contact between the flat end punch and flat plate (left) is defined as a Node to Node contact. The contact between the spherical punch and flat plate (right) is defined as a Surface contact.

Notice that the Surface contact is more general, but less numerically efficient, than the Node to Node contact. Every Node to Node contact could be defined as a Surface condition, but that would unnecessarily complicate the model.

Procedure

In SolidWorks, open the file named HANGER ASSEMBLY, shown in figure 9-4, and move it into COSMOSWorks.

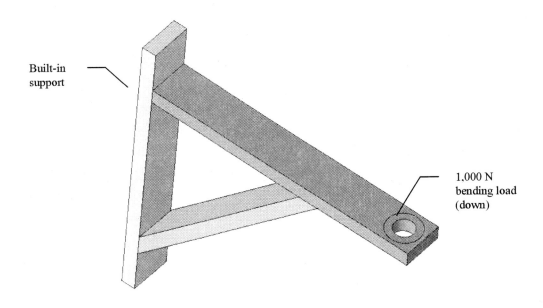

Figure 9-4: Hanger assembly

The hanger assembly consists of three parts (compare with figure 9-6). 1,000 N bending load is applied to the split face; support is applied to the back of vertical flat.

If you do not modify the Contact/Gaps parameters, then by default all touching faces are bonded, and the FEA model will behave as one part. A sample result is shown in figure 9-5.

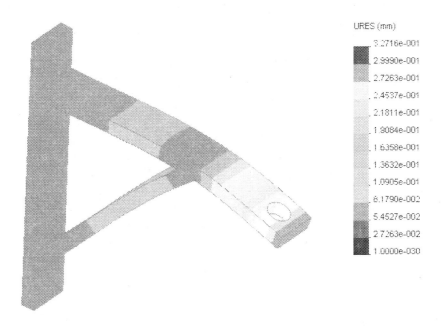

Figure 9-5: Displacement results for a model with all touching faces bonded

Notice that hanger assembly model is adequate for analysis of displacements. However, due to its sharp re-entrant corners, it is not suitable for analysis of maximum stresses, which are singular in all sharp re-entrant corners.

Now, we will modify the Contact/Gaps conditions on selected touching faces. We will leave the global conditions set to **Touching Faces: bonded**, but locally we will override them by defining a local **Contact/Gaps** condition. One of three pairs of touching faces will be defined as **Touching Faces: Free**, meaning that there is no interaction between faces. The faces, shown in figure 9-6, will be able to either came apart or "penetrate" each other with no consequences.

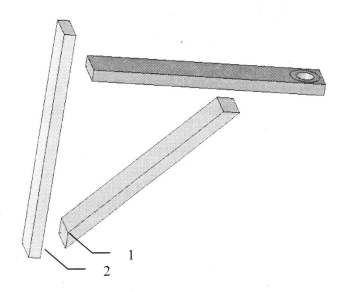

Figure 9-6: Touching Faces: Free

*When faces 1 and 2 are defined as **Touching Faces: Free**, there is no interaction between them.*

Note that exploded view of the hanger makes it easier to define local Contact/Gaps conditions.

Every time Contact/Gaps conditions are changed COSMOSWorks prompts you to remesh (figure 9-7).

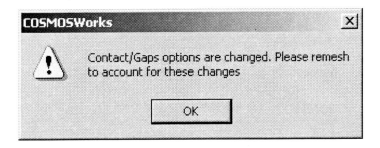

Figure 9-7: COSMOSWorks prompt when Contact/Gaps conditions change

Any change in Contact/Gaps conditions requires remeshing.

Remeshing will delete the previous results. If we wish to keep the earlier results, we need to redefine subsequent Contact/Gaps conditions in a new study.

Once local or component Contact/Gaps conditions have been defined, the *Contact/Gaps* icon becomes a folder, as shown in figure 9-8.

Figure 9-8: *Contact/Gaps* folder

The Contact/Gaps *icon becomes a folder once local or component conditions have been defined. Also shown is the pop up menu used to suppress, edit or delete Contact/Gaps conditions. Right-clicking the Contact Pair icon opens this pop-up menu.*

The lack of interference between faces in a pair defined as **Touching Faces: Free** is best demonstrated by presenting displacement results (figure 9-9).

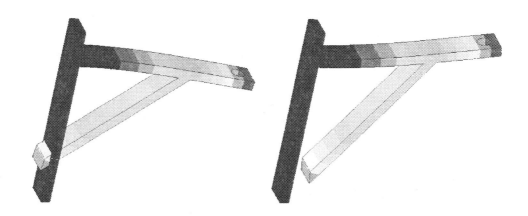

Figure 9-9: Displacement results in a pair defined as **Free**

The plot on the left shows results for load directed downwards; the plot on the right has the load direction reversed.

To change the local condition, right-click the *Contact Pair1* icon in *Contact/Gaps* folder (figure 9-8). This action opens a pop-up menu from which you select **Edit definition…** to open the Contact Pair window (figure 9-10).

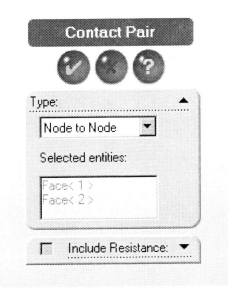

Figure 9-10: Contact pair window

The Contact pair window allows you to change local Contact/Gaps conditions.

In the Contact Pair window, change the local condition to **Node to Node**. Remesh the model when prompted, and run the solution once again. Notice that the solution now requires much more time to run because the contact constraints must be resolved (figure 9-11). Node to Node conditions, as well as Surface conditions, represent nonlinear problems and require an iterative solution, which often takes a long time.

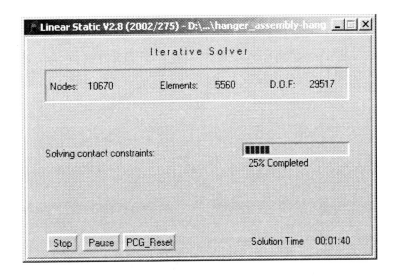

Figure 9-11: Iterative solution for the hanger assembly using a Node to Node condition

A Node to Node condition requires an iterative solution to solve contact constraints and takes significantly longer to complete than a linear solution.

The displacement results, shown in figure 9-12, show that the two faces defined as **Node to Node** now slide when the load points down (left) and separate when the load points up (right).

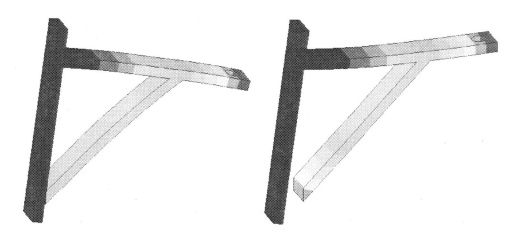

Figure 9-12: Displacement results for two locally defined Node to Node faces

*The two faces in the pair defined locally as **Node to Node** slide (left) or separate (right) depending on the load direction.*

Closer examination of the displacement results for the sliding faces (figure 9-13) shows that the sliding faces partially separate.

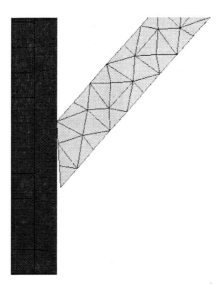

Figure 9-13: Partial separation of the sliding faces (figure 9-12, left)

Only a portion of face 1 contacts face 2. The mesh in the contact area is too coarse to allow for the analysis of contact stresses.

Note that while the mesh is adequate for the analysis of displacements, the mesh is not sufficiently refined for the analysis of the contact stresses that develop between the two sliding faces.

NOTES:

10: Analysis of two cylinders in contact

Objectives

On completion of this exercise, you will be able to:

- Perform an assembly analysis with surface contact conditions
- Use proper restraints to avoid rigid body modes

Project description

We will perform one more contact stress analysis. This analysis requires the Surface type of Contact/Gaps conditions. The analysis model, shown in figure 10-1, consists of two identical plates in contact through cylindrical surfaces. The material for both is Nylon 6/10. Our objective is to find the distribution of von Mises stress and the maximum contact stresses that develop in the model under 1,000 N of compressive load. The model geometry located in the SolidWorks assembly file is called TWO CYLINDERS.

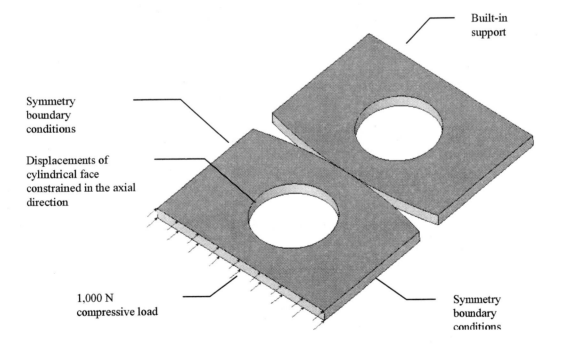

Figure 10-1: Two plates with their cylindrical surfaces in contact

Procedure

Preparation of the model for analysis requires constraining the loaded part to prevent rigid body motion and, at the same time, make it free to move in the direction of the load. We'll explain the required restraints that are shown in figure 10-1. Symmetry boundary conditions are applied to both side faces of the loaded part to act as guides. Displacements of the cylindrical face of the round hole are constrained in an axial direction in order to prevent rigid body motion in the direction defined by the axis of the hole. Constraining the loaded part in this direction is required because the analyzed contact is frictionless. Actually, we could constrain any one point of the loaded part (for example, a corner) in the same direction.

Restraint windows, where you define the restraints to the loaded part, are shown in figure 10-2.

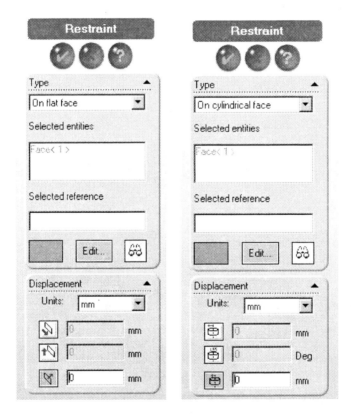

Figure 10-2: Restraint windows

These windows display the definition of symmetry boundary conditions on side face of the movable part (left) and the definition of the restraints preventing the "up and down" movement of the loaded part of the assembly (right).

We are now ready to mesh the assembly model. Adequate mesh density in contact area is of paramount importance in any contact stress analysis. It is the responsibility of the user to make sure that there are enough elements in the contact area to properly model to the distribution of contact stress.

To investigate the impact of mesh density in the contact area on the contact stress results, we will run the analysis twice; first with the default mesh density as suggested by the automesher, then with the mesh density set twice as high (figure 10-3).

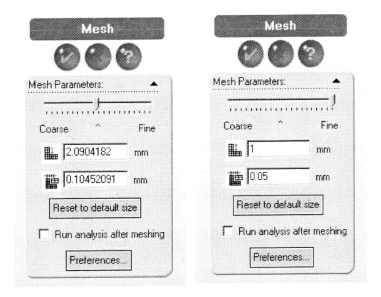

Figure 10-3: Mesh controls used to create a coarse mesh (left) and a fine mesh (right)

Contact stress results for the coarse and fine meshes are shown in figures 10-4 and 10-5 respectively. Note that in order to analyze contact stress, the third principal stresses P3 are displayed. Because contact stresses are compressive, the maximum contact stress is shown at the opposite end of the color scale in blue.

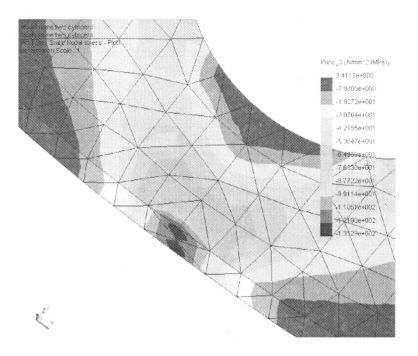

Figure 10-4: Contact stress results for the coarse mesh

The maximum contact stress reported by the coarse mesh is 132 MPa.

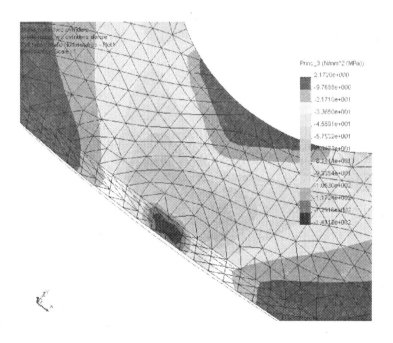

Figure 10-5: Contact stress results for the fine mesh

The maximum contact stress reported by the fine mesh is 141 MPa.

The stress results produced by two meshes are quite close and prove that even the first mesh was adequate. Note that the required mesh size depends on the material. We conveniently used Nylon 6/10 material, which has a low modulus of elasticity. Due to this soft material, deformations, and consequently the contact area, are large and this condition makes large elements permissible.

We leave it to the reader to decide if the stresses are acceptable for this nylon material (look at stresses everywhere in the model, not just in the contact area).

Von Mises stress results that were produced by the denser mesh are shown in figure 10-6.

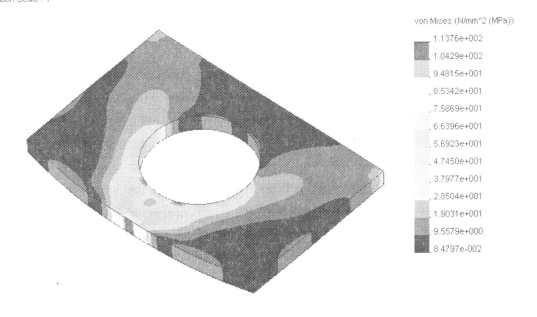

Figure 10-6: Von Mises stress results obtained using the denser mesh

Please further investigate the effects of mesh refinement on contact stress results when using different materials, such as steel or aluminum.

NOTES:

11: Analysis of a bi-metal beam

Objectives

On completion of this exercise, you will be able to:

❑ Conduct a thermal stress analysis

❑ Use various techniques in defining restraints

Project description

The temperature of the bi-metal beam, shown in figure 11-1, increases uniformly from an initial 20°C to 300°C. We need to find how much the beam will deform.

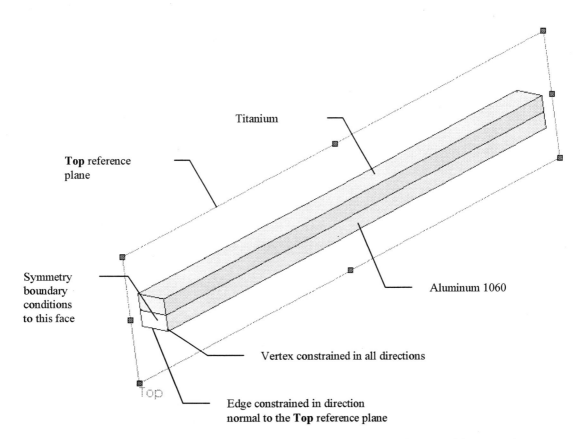

Figure 11-1: Bi-metal beam and the Top reference plane used in Restraint definition

*The bi-metal beam will deform when heated because of the different thermal expansion ratios of titanium and aluminum. The **Top** reference plane used in the **Restraint** definition is also shown in this diagram.*

Procedure

An important part of this exercise is applying constraints in such a way as to allow for deformations due to thermal stress, but preventing rigid body motions. In other words, we need to restrain the model as little as possible. We can is approximate this effect in several different ways. For example, we could apply symmetry boundary conditions to one of the end faces, constrain one edge of that face in the direction normal to that edge, and constrain one vertex of that face in all directions. We say "approximated" because restraints will keep the affected face planar and the edge straight. The details of **Restraints** definition are presented in figure 11-2.

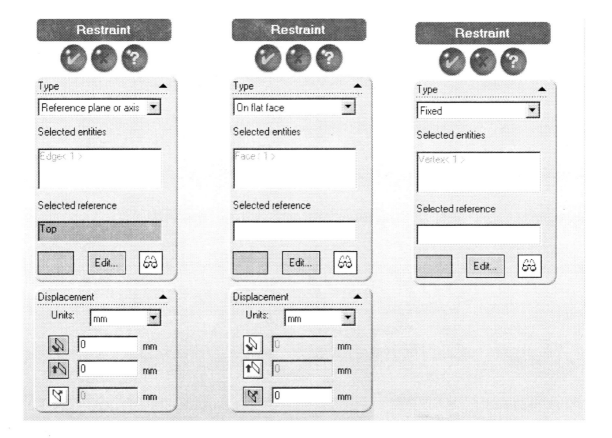

Figure 11-2: Restraint windows showing restraints applied to the edge (left), to the face (middle) and to the vertex (right)

*Note that edge restraint definition uses the **Top** reference plane from SolidWorks model for reference.*

To account for thermal effects, we define the study as *Static*, but in the Study window, we select the option **Include thermal effects** (figure 11-3). There is no need to run a thermal analysis because thermal effects are accounted for in the static study, which is of structural, not thermal type.

Engineering Analysis with COSMOSWorks

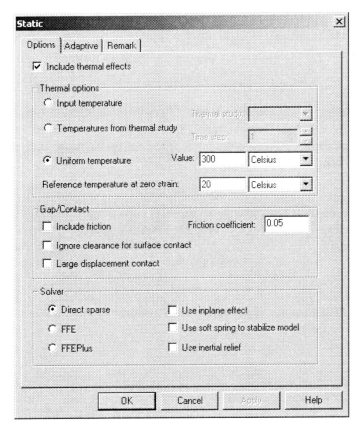

Figure 11-3: Study window, where we request that thermal effects be included.

*To consider thermal effects, we select the option, **Include thermal effects** under the Options tab. In the Thermal options area, we define the uniform temperature to which the model is heated and the Reference temperature at zero strain. Using both temperatures, COSMOSWorks calculates the increase (or decrease) of temperatures to which the model is subjected.*

Before proceeding, let's take this opportunity to review all thermal options available in the Study window.

Thermal Option	Definition
Input temperature	Use if prescribed temperatures will be defined in the *Load/Restraint* folder of the study to calculate thermal stresses.
Temperature from thermal study	Use if temperature results are available from previously conducted thermal study.
Uniform temperature	Use to calculate thermal stresses with steady temperature change throughout the model. If uniform temperature is selected, the program ignores all prescribed temperature definitions.
Reference temperature at zero strain	Use to specify the temperature at zero strain.

Note that **Reference temperature at zero strain** (Figure 11-3) lets us specify the temperature at zero strain. In this exercise we could either:

- Define the Input Temperature for both assembly components, or
- Use the **Uniform temperature** option in Study properties.

Using **Uniform temperature** is more straightforward because we don't have to define any temperature loads; therefore, we will use this method for the exercise. However, are encouraged to run the same analysis using **Input Temperature**.

Now let's assign the material properties of titanium and aluminum to the assembly components (figure 11-1) and mesh the model with an element size of 0.75 mm. This element size assures eight layers across the assembly with each strip 3 mm thick. We need this many elements to examine stress results. For a deformation analysis, eight layers would definitely be excessive as two layers would suffice.

Displacement results are shown in figure 11-4, and stress results are shown in figure 11-5.

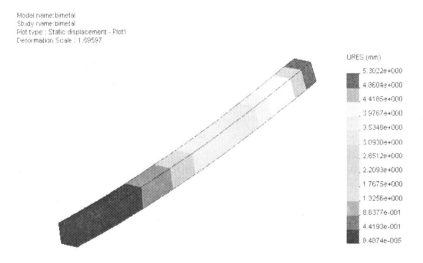

Figure 11-4: Displacement results for the bi-metal beam

Due to the different thermal expansion ratios of titanium and aluminum, thermal strains develop and deform the bimetal beam in a pattern that resembles bending.

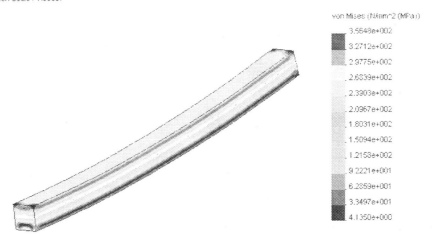

Figure 11-5: Stress results for the bi-metal beam

Thermal stresses do not follow the stress pattern typical for bending.

NOTES:

12: Analysis of a pipe with cooling fins

Objective

On completion of this exercise, you will be able to:

❑ Perform a transient thermal analysis

Project description

We will analyze an aluminum pipe with cooling fins. At a point in time, t = 0 s, temperature of 80°C is applied to the inside surface of the pipe. We wish to find how the temperature on the external faces changes with time and determine the equilibrium temperature. All required input data are shown in figure 12-1. Use the SolidWorks model in the file named PIPE WITH COOLING FINS.

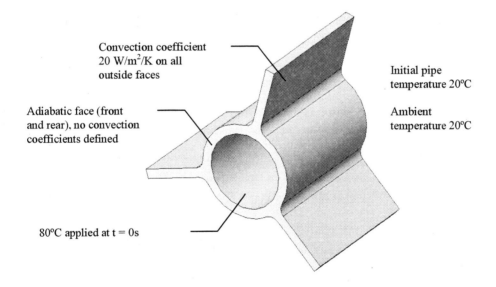

Figure 12-1: Aluminum pipe with cooling fins

This aluminum pipe with three cooling fins has a prescribed temperature 80°C applied to the internal face. This temperature does not change with time.

Procedure

The problem, how long it takes for the temperatures to stabilize, requires a transient thermal analysis. Specify this type of analysis by selecting **Transient** as the Solution type under the Options tab in the Thermal window, where you define the properties of a thermal study (figure12-2).

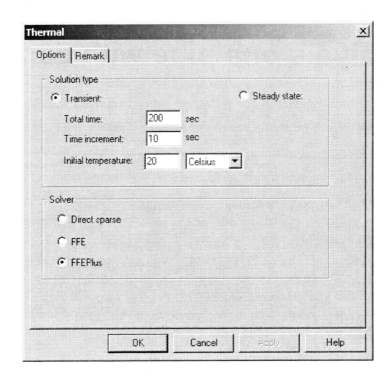

Figure 12-2: Selected options in the Thermal window

*Either **Steady state** or **Transient** can be selected as the thermal analysis type. The initial temperature of the model is a uniform 20°C.*

As shown in figure 12-2, the total analysis time is specified as 200 s, with measurement points every 10 s. This frequency of measurement captures 20 solution steps in 5 s time intervals. Because there is no assurance that temperatures will stabilize towards the end of the time period, the specified duration of analysis (200 s) is a guess.

The windows for defining the prescribed temperature and convection coefficients are shown in figure 12-3. Note that in our study, the ambient temperature specified in the convection coefficient definition window is 20°C, which is equal to the ambient temperature. This definition, however, is not a requirement, and different initial and ambient temperatures may be specified if desired.

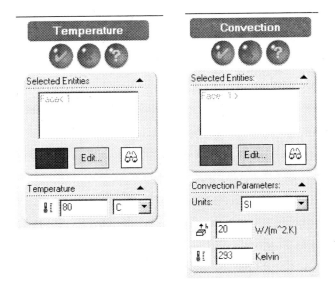

Figure 12-3: Definitions of prescribed temperature (left) and convection coefficients (right)

These values remain unchanged during the transient thermal analysis.

Sample results showing temperature distribution during the warm-up process are shown in figures 12-4 and 12-5.

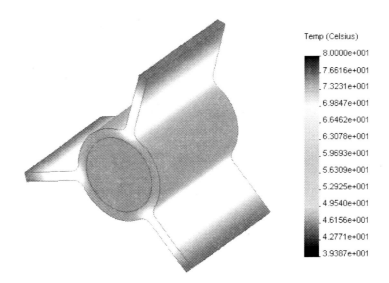

Figure 12-4: Temperature distribution after 20 seconds

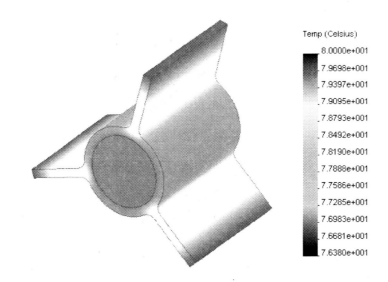

Figure 12-4: Temperature distribution after 200 seconds

The last calculated time step is at 200 s.

Simple examination of the temperatures corresponding to the calculated step, as shown in figure 12-4, is not sufficient to conclude that the last calculated step represents the equilibrium temperature. Remember that we chose the duration of 200 s with no assurance that temperatures would stabilize in those 200 s. Therefore, let's create ten plots showing the temperatures every 20 seconds and plot the temperature of the tip of the cooling fin as a function of time (figure 12-5).

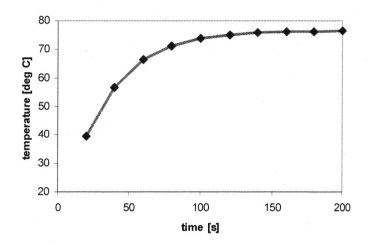

Figure 12-5: Temperature of the tip of cooling fin as a function of time.

Examination of the plot in figure 12-5 proves that equilibrium temperature has been effectively reached after 200 s.

NOTES:

13: Analysis of an L-beam

Objectives

On completion of this exercise, you will be able to:

- Perform a buckling analysis
- Determine the buckling load factor

Project description

An L-shaped beam is compressed with 50,000 N force, as shown in figure 13-1. The beam material is alloy steel with a yield strength of 620 MPa. Our goal is to calculate the factor of safety. Open the model geometry in the SolidWorks part file called L BAR.

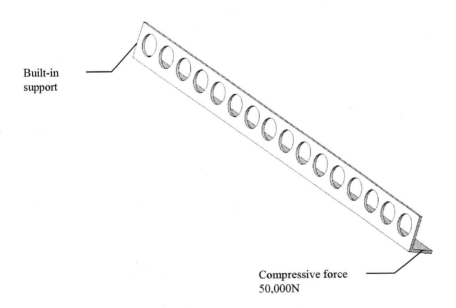

Figure 13-1: L-beam geometry

A perforated channel is compressed by a 50,00 N force uniformly distributed over the end face.

Procedure

We'll now present another type of analysis to complement the static, frequency and thermal analyses discussed in previous chapters. In this chapter we'll perform a buckling analysis.

Before we run a buckling analysis, let's first obtain the results of a static analysis based on the load and restraint shown in figure 13-1. The results of the static analysis show the minimum factor equal to 2.3 (figure 13-2). Figure 13-3 identifies the location of the highest stress.

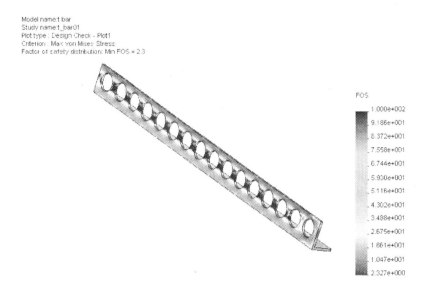

Figure 13-2: Distribution of the factor of safety

The lowest factor of safety is 2.3.

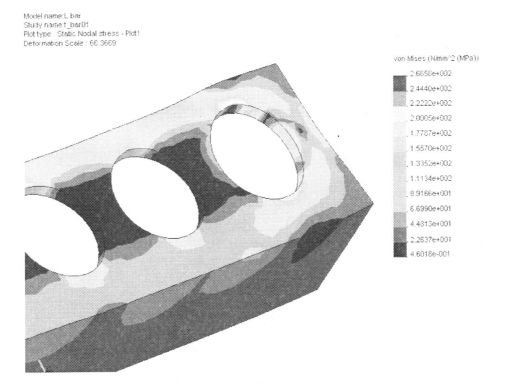

Figure 13-3: Location of maximum von Mises stress (in the hole closest to the loaded end)

As always the case with slender members under compressive load, the factor of safety, related to material yield, is not sufficient to make conclusions about the structure's safety. Because of the possibility of buckling, we still need to calculate the safety factor for a buckling load, and this calculation requires running a buckling analysis. Figure 13-4 shows two studies: *l_bar01* is a static analysis, *l_bar02* is a buckling analysis.

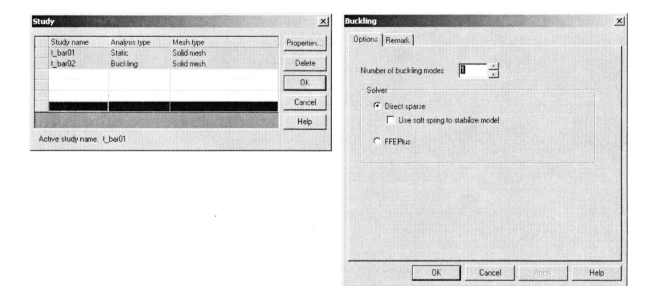

Figure 13-4: Definition of a buckling study (left) and the Buckling window, where properties are defined for a buckling study

Defining a buckling study requires specifying the number of desired buckling modes.

When defining a buckling study, we need to decide how many buckling modes should be calculated. This is in close analogy to the number of modes in a frequency analysis. In most practical cases, the first buckling mode determines the safety of the analyzed structure. Therefore, we will limit our analysis to calculating the first buckling mode. Once the buckling analysis has been run, COSMOSWorks automatically creates two *Results* folders: *Displacement* and *Deformation*. Even though, as shown in figure 13-5, displacement results are available, they do not provide much useful information. In a buckling analysis, the magnitude of displacement results is meaningless, just like the results in a modal analysis. The deformation results are much more informative (figure 13-6).

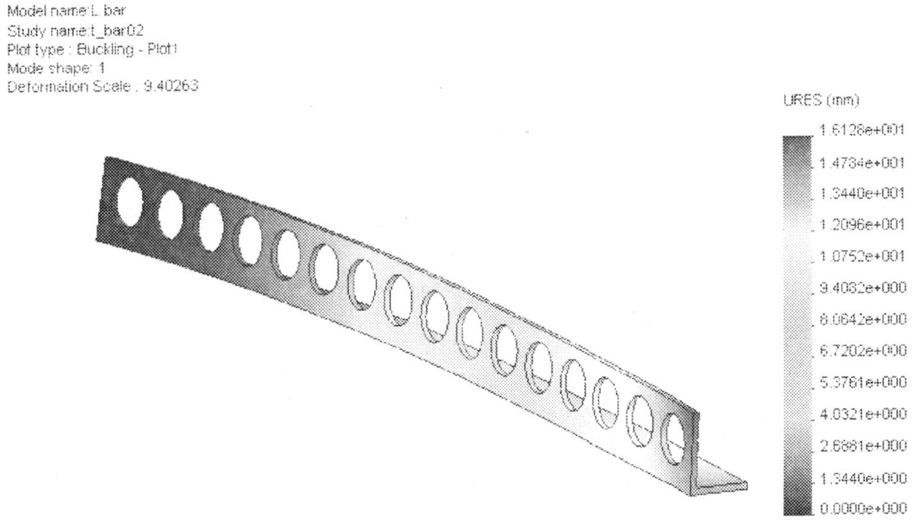

Figure 13-5: Displacement magnitudes in a buckling analysis

Displacement magnitudes are meaningless in a buckling analysis. Therefore, displacement results can only be used to compare relative displacements between different portions of the model in the given buckling mode.

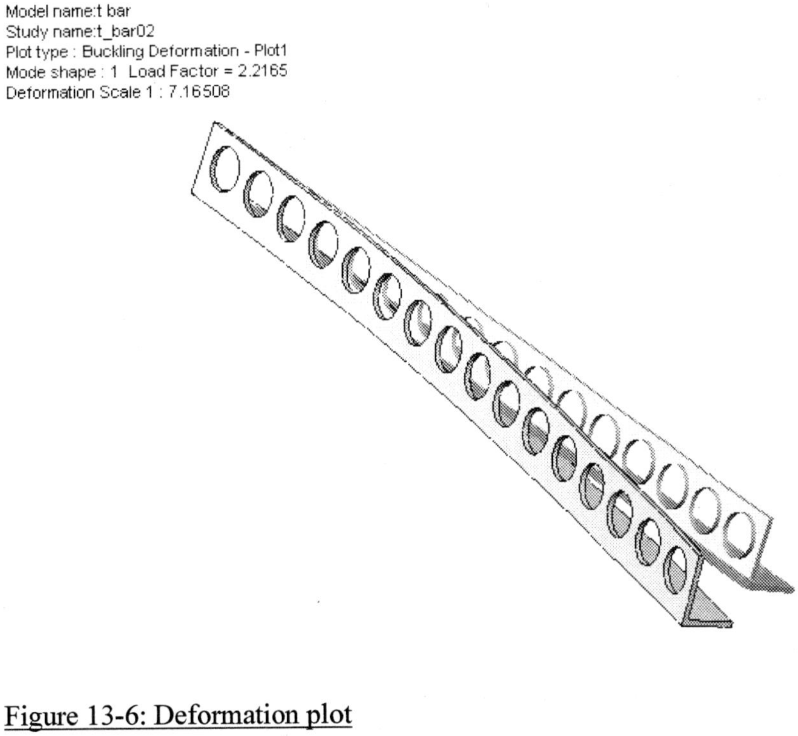

Figure 13-6: Deformation plot

The deformation plot provides visual information on the shape of the buckled structure. It also lists the buckling factor, here equal to 2.2. The plot shows the buckled shape along with the undeformed model.

The buckling load factor, shown in the deformation plot, provides information on by how many times the load magnitude would need to be increased in order for buckling to take place. In our case, the magnitude of load causing buckling is 2.2165*50,000 N = 110,825 N. The buckling load factor, therefore, doubles as the buckling-load safety factor. Notice that the calculated buckling-load safety factor is actually lower than the previously calculated safety factor related to material yield strength. This condition means that the beam will buckle before it develops stresses equal to the yield strength. We, therefore, conclude that buckling is the deciding mode of failure. Also notice that high stress affects the beam only locally, while buckling is global.

We should point out that the calculated value of the buckling load is non-conservative. It does not account for the always-present imperfections in model geometry, materials, loads, and supports. The real buckling load may, therefore, be significantly lower than the calculated 110,825 N.

14: Optimization of a plate in bending

Objectives

On completion of this exercise, you will be able to:

- Explain factors in an optimization analysis: objective, design variables, constraints
- Perform an optimization analysis

Project description

A rectangular plate, shown in figure 14-1, is subjected to 500 N bending load resisted by built-in support. We suspect that the plate is over-designed and wish to find out if material can be saved by enlarging the hole diameter. Let's say that because of certain design considerations, the diameter cannot exceed 80 mm. Also, from previous experience with similar structures, we know that the highest von Mises stress should not exceed 500 MPa anywhere in the plate. The model geometry, called PLATE IN BENDING, can be found in the SolidWorks part file. Use alloy steel for the material.

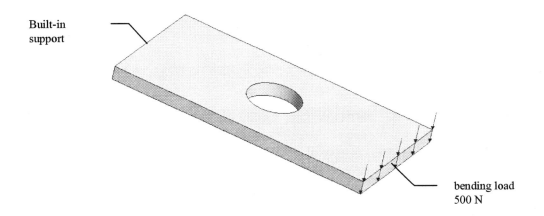

Figure 14-1: Rectangular plate with a round hole is bent by 500 N load.

Procedure

What we are facing here is a design optimization problem. As with every design optimization problem, this one too is defined by the optimization goal, design variables, and constraints. Before proceeding we need to explain those terms.

Term	Definition
Optimization goal	The objective is to minimize the mass. Other examples of optimization objectives are to maximize stiffness, maximize the first natural frequency, etc. The optimization goal is often called the optimization objective or optimization criterion.
Design variable	The allowed range of change of some parameter always characterizes design variables. In our case, the hole diameter is the design variable. For simplicity, this exercise has only one design variable, but there is no limit on the number of design variables. In this problem, the range is defined by the hole diameter changing from its current value of 40 mm to the maximum allowed value of 80 mm.
Constraints	For example, limits on the maximum deflection or the minimum natural frequency are examples of constraints. Constraints in optimization exercises are also called limits. In this exercise, constraint is the maximum von Mises stress, which must not exceed 500 MPa.

Before proceeding with the design optimization, which will result in changing the hole diameter, it is necessary to determine the stresses in the plate "as is". Determining these stresses requires running a static study, which is a prerequisite to the subsequent optimization study.

Figure 14-2 shows model dimension before optimization, and figure 14-3 shows the mesh created with the default element size. Notice that the mesh has two layers of second order elements across the member in bending, as is recommended for bending problems. Von Mises stress results are presented in figure 14-4.

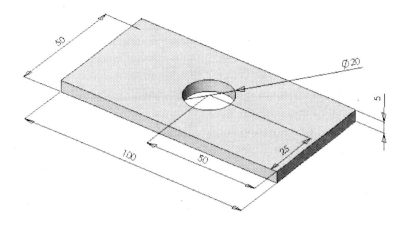

Figure 14-2: Model dimensions before optimization

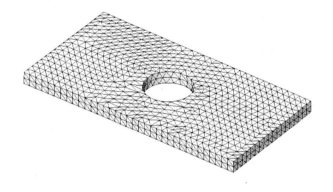

Figure 14-3: Mesh of the model before optimization

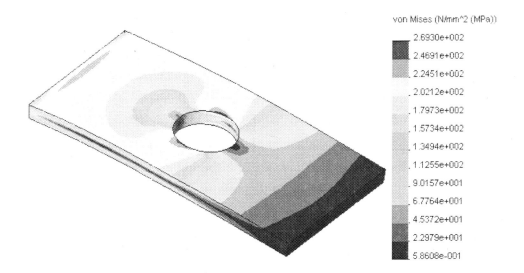

figure 14-4: Von Mises stresses in the model before optimization

Stresses shown in figure 14-4 report the maximum von Mises stress of 269 MPa at the edge of the hole. Since we allow 500 MPa, there is room for saving material by making this hole bigger, so we can proceed with the optimization exercise.

The optimization study is defined as any other study except for the mesh type (figure 14-5). Mesh type does not need to be defined because the optimization study uses the same type of mesh as the prerequisite static study.

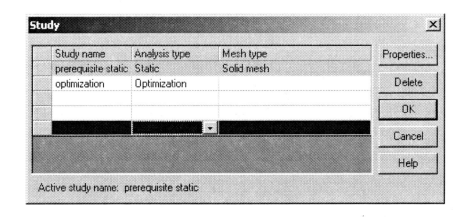

Figure 14-5: Optimization study and static study

The optimization study, called *optimization*, is defined together with the static study, called *prerequisite static*, which is, as its name implies, a prerequisite to the optimization study.

When the optimization study is defined, COSMOSWorks automatically creates three folders specific to an optimization study (figure 14-6):

- *Objective*
- *Design Variables*
- *Constraints*

An optimization study requires that all these folders be defined. Design optimization is an iterative process. The maximum number of design cycles is defined in the Optimization window, which you open by right-clicking the *Optimization* study, and from there, you can select the number of iterative cycles (figure 14-7).

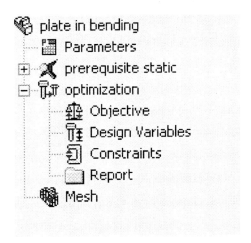

Figure 14-6: Design optimization study contains three automatically created folders: *Objective*, *Design Variables*, and *Constraints*.

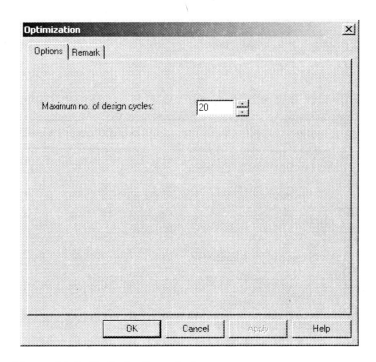

Figure 14-7: Optimization window

The maximum allowed number of design cycles in an optimization study is 20 by default.

The optimization goals are defined in the Objective window. To open this window, right-click the *Objective* folder to open the Objective window, shown in figure 14-8. For this exercise, we accept the default optimization goal, which is to minimize the mass (figure 14-8).

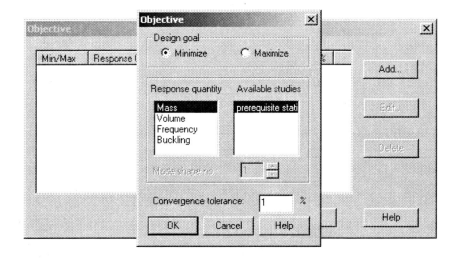

Figure 14-8: Optimization goal, defined in the Objective window, to minimize mass

The prerequisite analysis, here called prerequisite static, *is also selected in this window.*

To define the design variable, first select the dimension, which will be modified by this design variable. You can display the dimensions by selecting **Show Feature Dimensions** in *Annotations* folder in the SolidWorks Manager. To define the design variable, right-click the *Design Variable* folder to open the Design Variable window, shown in figure 14-9. The allowed range of variation of hole diameter is from 20 mm to 40 mm. Note that it may not be possible to reach the diameter of 40 mm if, during the process of increasing the diameter, the maximum stress exceeds 500 MPa.

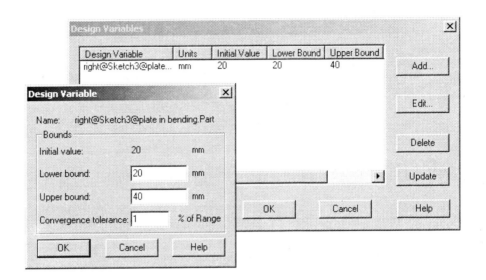

Figure 14-9: Design Variable window

The allowed range of variation of hole diameter is specified as from 20 mm (value before optimization) to 40 mm, which is the maximum allowed hole diameter.

Finally, to define the constraints, right-click the *Constraints* folder to open Constraint window, shown in figure 14-10.

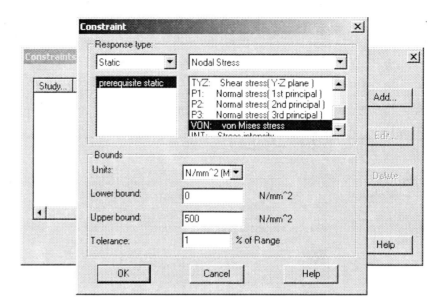

Figure 14-10: Constraint window

The maximum allowed stress is defined as von Mises stress equal to 500 MPa. The prerequisite static study is also specified in this window.

Notice that loads, supports, and materials are not defined anywhere in the optimization study. Also notice that no mesh needs to be created. The necessary information is transferred from the prerequisite study defined in Constraint window.

Having defined the optimization goal, design variable(s), and constraint(s), we will now run the optimization study. Optimization is an iterative process, during which the value of the design variable(s) is changed within the specified range and each change requires a new mesh. When the solution is complete, COSMOSWorks creates several *Results* folders, which are shown in figure 14-11.

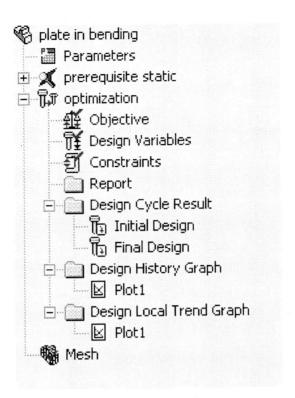

Figure 14-11: COSMOSWorks *Results* folders following optimization

The following folders are automatically created once design optimization completes:

- *Design Cycle Result*
- *Design History Graph*
- *Design Local Trend Graph*

To view the optimized model, double-click the *Final Design* icon in the *Design Cycle Result* folder (figure 14-12). To view the original model for comparison, double-click the *Initial Design* icon in the *Design Cycle Result* folder (figure 14-2).

Engineering Analysis with COSMOSWorks

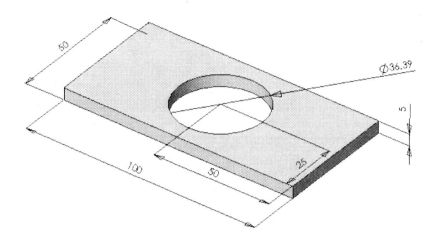

Figure 14-12: Model after optimization

In the optimized model, the hole diameter is 36.39 mm.

If desired, we can display the model configuration in any step of the iterative design optimization process (figure 14-13).

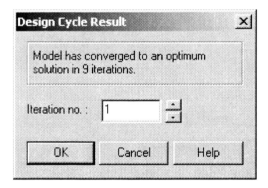

Figure 14-13: Design Cycle Result window

Open the Design Cycle Result window by right-clicking the *Design Cycle Result* folder. The Design Cycle Result window allows you to display the shape of the model shape at any point during the iterative optimization process.

To examine displacements or stresses in the optimized model, we need to review the plots in the prerequisite static study, the results of which have been updated to account for the new model geometry.

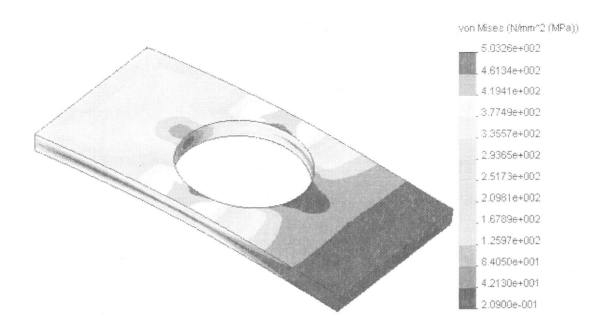

Figure 14-4: Von Mises stresses in the optimized model

Note that the maximum stress is 503 MPa. This value is within the requested 1% accuracy of constraint.

The history of the optimization process can be reviewed by examining plots in the *Design History Graph* folder and in the *Design Local Trend Graph* folder. An example is shown in figure 14-5.

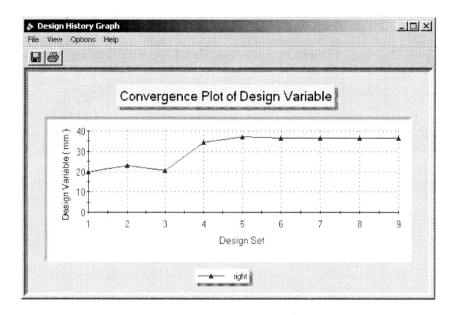

Figure 14-5: Graph showing changes in the design variable during the iterative optimization process

15: Analysis of a hollow bracket

Objectives

On completion of this exercise, you will be able to:

❑ Use the p-Adaptive solution method

❑ Compare between h-elements and p-elements

Project description

A hollow cantilever bracket, shown in figure 15-1, is supported along the back side. A bending load of 10,000 N is uniformly distributed to the cylindrical face. We need to find the location of the maximum von Mises stress and calculate its magnitude. Model geometry can be found in the SolidWorks parts file, called HOLLOW BRACKET.

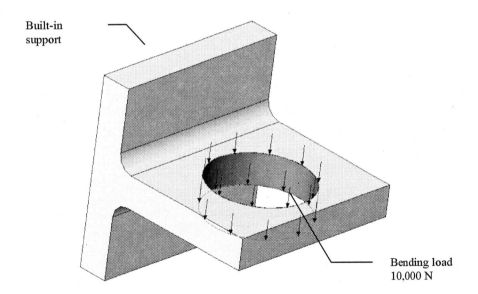

Figure 15-1: Hollow cantilever bracket under a bending load

Notice that the bracket geometry has been defeatured; there are no external rounds and no details. Due to the symmetry of the bracket geometry, loads, and supports, we could simplify the geometry further by cutting it in half, but decide that the efforts in geometry idealization are not worth the gains.

Another reason for not simplifying the geometry is that you are encouraged to use the same geometry later to perform a frequency analysis.

Procedure

The presented problem is a straightforward structural analysis and hardly seems to deserve its place towards the end of this book. However, we'll use this problem to introduce a whole new concept in FEA; we will solve this problem using a different type of finite elements, called p-elements. Before we begin, we need to explain what p-elements are and what they do.

If you recall, in chapter 1, we said that COSMOSWorks uses two types of elements: tetrahedral solids and triangular shells. Each can be defined as either:

❏ First order element (draft quality)

❏ Second order element (high quality)

Further recall that first order elements model a linear (or first order) displacement and constant stress distribution, while second order elements model a parabolic (second order) displacement and linear stress distribution.

We also said that even though second order elements are more computationally demanding, they are better able to model stress gradients and map well to curvilinear geometry. Therefore, second order elements have definite advantage over first order elements.

We now have to amend the above paragraphs. Beside first and second order elements, COSMOSWorks also has higher order elements, with orders up to the 5th. You can access those higher order elements, or p-elements, if **Use p-Adaptive for solution** is selected (checked) in the Study window under the Adaptive tab (figure 15-2). This option is available only for static analysis when using solid elements.

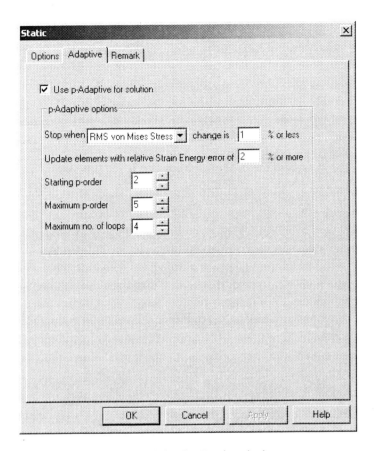

Figure 15-2: Adaptive tab in the Static window

The Adaptive tab in the Static window allows you to use the p-Adaptive solution method.

If we use settings shown in figure 15-2 used, when you select **Use P-Adaptive for solution**, you can define the other p-Adaptive options. **Starting p-order** is set to 2, which means that all elements are defined as second order elements. The **p-Adaptive solution** runs in iterations, called loops, and with each new loop, the order of elements is upgraded. The highest order available is the 5th order, but the highest order to be used is defined by **Maximum p-order** (set to 5 in figure 15-2). The **Maximum number of loops** is set to 4. Looping continues until the change in Root Mean Square (RMS) von Mises stress between the two consecutive iterations is less than 1%, as specified in the p-Adaptive options area. If this requirement is not be satisfied, then looping will stop when the elements reach the 5th order; this will be 4th loop.

RMS von Mises stress calculates one mean stress value for the whole model and is used here as a convergence criterion because changes in its value control the iteration process. Please investigate the available options in the other p-Adaptive options fields.

Note that elements used in p-Adaptive solutions do not have a fixed order, but can be upgraded "on the fly", that is, automatically during the iterative solution process without our intervention. These types of elements with upgradeable order are called p-elements. The p-Adaptive solution is close in analogy to the iterative process of mesh refinement, which also continues until the change in the selected result is no longer significant.

Let's pause for a moment and explain some terminology:

Why are upgradeable elements called p-elements?

Let's go back to chapter 2. As figure 2-14 explains h denotes the characteristic element size. This size is manipulated during the mesh refinement process. While mesh is refined, the characteristic element size, h, becomes smaller. Therefore, the mesh refinement process that we conducted in chapters 2 and 3 is called the h-convergence process, and the elements used in this process are called h-elements. Note that h-elements retain their order. Once created, they cannot be upgraded to a higher order.

The iterative process, that we are discussing now, does not involve mesh refinement. While mesh remains unchanged, the element order changes from the 2nd all the way to 5th (or less if the convergence criterion is satisfied sooner).

The element order is defined by the order of polynomial functions that describe the displacement field in the element. Because the polynomial order experiences changes, the process is called p-convergence process, and the upgradeable elements used in this process are called p-elements.

Why is the p-convergence process called a p-Adaptive solution, and what exactly does "Adaptive" mean?

Adaptive means that not necessarily all p-elements need to be upgraded during the solution process. Indeed, as you see in figure 15-2, in the p-Adaptive options area, the field in **Update elements with relative Strain Energy error of ___ % or more** is set to **2**, meaning that only those elements not satisfying the above criterion will be upgraded (please investigate other criteria). We say, therefore, that element upgrading is "Adaptive", or driven by the results of consecutive iterations.

We are now sufficiently familiarized with p-elements to proceed with the exercise.

Procedure, continued

First we'll solve our model the way we have always done it, using second order solid tetrahedral h-elements. We use the default element size, but to better capture stresses, we apply **Mesh Control** to both rounds (figure 15-3) by selecting **Automatic transition** in the Preferences window under the Mesh tab. The h-element mesh is shown in figure 15-4 and the von Mises stress results in figure 15-5.

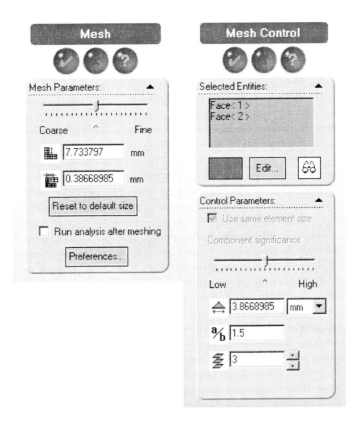

Figure 15-3: Mesh window (left) and Mesh Control window (right) used to create h-element mesh

The Mesh window (left) displays the Mesh Parameters. The Mesh Control window (right) displays the controls, applied to both rounds, which are used to create an h-element mesh.

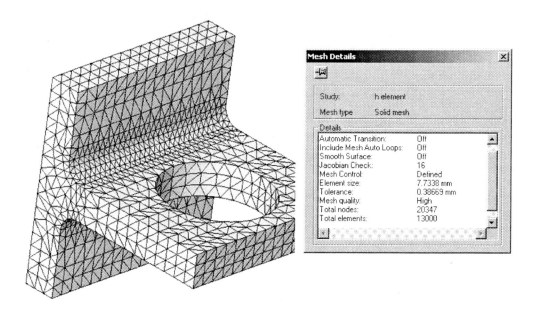

Figure 15-4: h-element mesh of second order tetrahedral elements

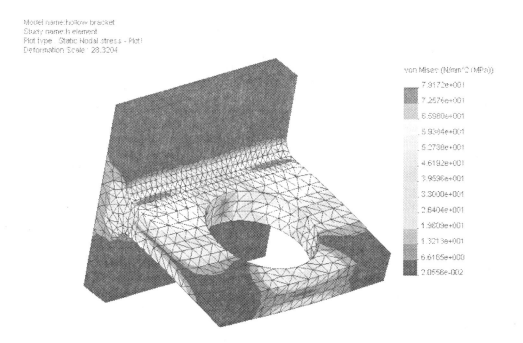

Figure 15-5: Von Mises stress results obtained using h-elements

The maximum stress is 79.16 MPa.

Now let's create a new study identical the one we just finished analyzing results for, except that in the Study window under the Adaptive tab, we select **Use p-Adaptive for solution**. Restraints and Loads can be copied from the previous study.

Before meshing we make one observation. Considering that p-adaptive solution will be used, we can manage with a very coarse mesh, as shown in figure 15-6.

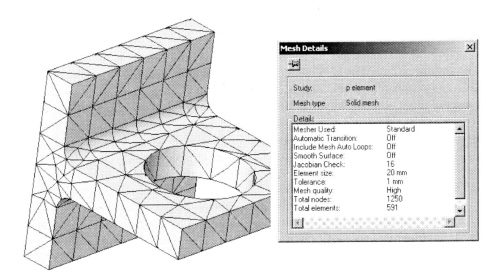

Figure 15-6: Mesh for use with p-Adaptive solution

The mesh shown in figure 15-6 would never be acceptable for use with h-elements. There would not be enough elements to capture the complex stress field near the rounds if h-elements were used. However, if we use higher order elements, which is equivalent to refinement of an h-element mesh, even this coarse mesh will deliver accurate results. Indeed, having solved the study with p-elements, we produce the stress plot shown in figure 15-7.

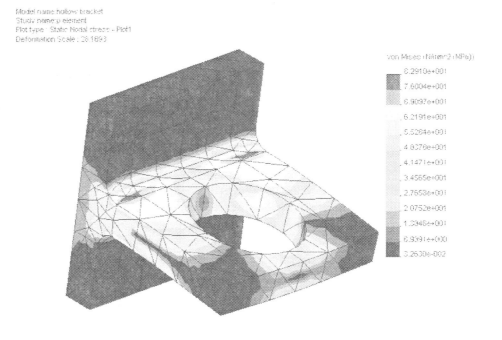

Figure 15-7: Von Mises stress results obtained using p-elements

The maximum stress is 82.91 MPa, very close to the 79.16 MPa obtained previously when using h-elements.

Once the **p-Adaptive solution** is ready, we can display the final result, as shown in figure 15-7. We can also access the history of the iterative solution. To do this, right-click the appropriate *Study* icon (study using p-elements) to open a pop-up window (figure 15-8, left) where you can select **Convergence Graph…**. The Convergence Graph window opens, from where you can specify what information to display on the graph (figure 15-8, right).

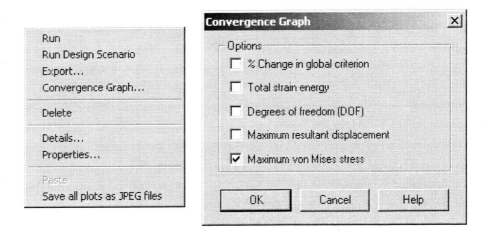

Figure 15-8: Pop-up window used to open the Convergence Graph window

To display the Convergence Graph window, right-click the appropriate Study *icon, which opens a pop-up window (left). In the pop-up menu, select* **Convergence Graph…**. *The Convergence Graph window (right) appears, where you can select display options for the convergence graph.*

Say we are primarily interested in how the accuracy of the maximum von Mises stress. We then select Maximum von Mises stress to be graphed throughout all performed iterations (figure 15-9).

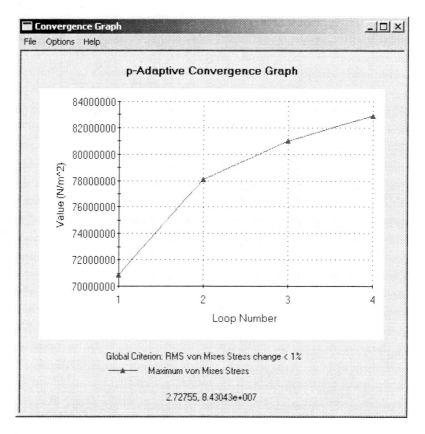

Figure 15-9: Graph showing maximum von Mises stress calculated in each loop of a p-Adaptive solution process

The graph also shows that four iterations were required to converge within 1% of RMS stress as specified in figure 15-2. The shape of the curve (convex) indicates that convergence is taking place.

Note that while the convergence process was controlled by RMS von Mises stress, which is a stress norm based on stress results in the entire model, the graph in figure 15-9 shows the maximum von Mises stress.

Which solution method is better: the "regular" method using h-elements or the p-Adaptive solution method presented in this chapter? Generally, with second order h-elements, we can obtain a reasonably accurate solution within a reasonably short time. Experience indicates that second order h-elements offer the best combination of accuracy and computational simplicity. For this reason, the automesher in COSMOSWorks is tuned to meet the requirements of an h-element mesh. The p-Adaptive solution method is much more computationally demanding and significantly more time-consuming. Therefore, the p-Adaptive solution method is reserved for special cases where, the solution must be known with narrowly specified accuracy. The p-Adaptive solution is also a great learning tool, leading to better understanding of element order, the convergence process, and discretization error. For this reason, readers are encouraged to repeat some of previous exercises using p-Adaptive solution method.

16: Analysis of a tapered block

Objective

On completion of this exercise, you will be able to:

❑ Perform a design sensitivity study using parameters and a design scenario

Project description

A tapered block is loaded with 10,000 N bending load uniformly distributed over the end face, as shown in figure 16-1. In the middle of its length, there is a hole running all the way through the block. We wish to investigate von Mises stress at locations 1, 2, and 3. We want to calculate the maximum von Mises stress at these three points and calculate the maximum resultant deflection while the distance between the support and hole center changes from 20 mm to 100 mm (figure 16-2). The model geometry, called TAPER, is in the SolidWorks part file.

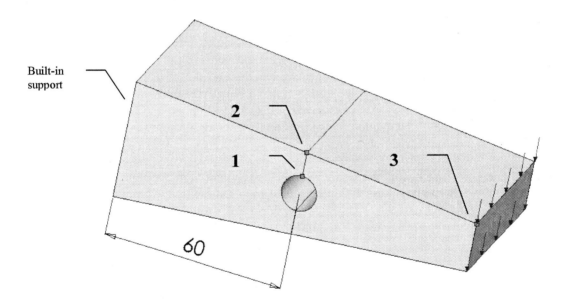

Figure 16-1: Locations 1, 2, and 3 on a tapered block

The numbered locations on the blocs are where we need to determine von Mises stress while the hole position is changed, as shown in figure 16-2.

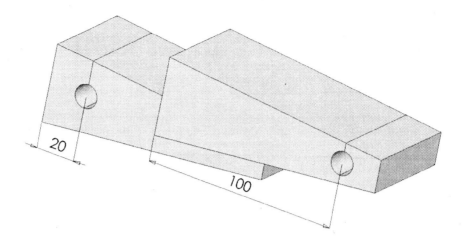

Figure 16-2: Initial hole location 20 mm away from the supported face (left) and final hole position 100 mm away from the supported face (right)

Procedure

Let's say that we wish to proceed by changing the hole position from 20 mm to 100 mm in 5-mm intervals. This would require running 17 analyses and a rather tedious compiling of all results. COSMOSWorks offers an easier way to accomplish our objectives. The distance defining the hole position can be defined as a Parameter and with one step, we can run a design scenario. The results of the design scenario can then be plotted using COSMOSWorks tools. A design scenario is often called a sensitivity study as it investigates the sensitivity of selected system responses (here, von Mises stress and the resultant displacement) to changes in certain parameters defining the model (here, the hole location).

We begin, as always, with the assignment of material properties (Aluminum 1060) and the definition of supports and loads. To define the parameters for this exercise, right-click the *Parameters* folder (figure 16-3) to open a pop-up menu and select **Edit/define…**.

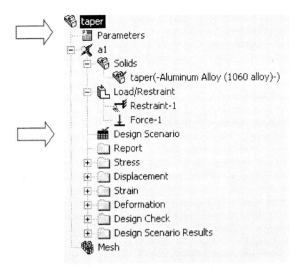

Figure 16-3: *Parameters* and *Design Scenario* folders

Parameters *and* Design Scenario *folders are created automatically and are used only when a Design Scenario is run. Notice that the* Parameters *folder is automatically created before any study is defined.*

Selecting **Edit/Define...** opens the Parameters window (figure 16-4, top). In the Parameters window, select **Add**, which opens the Add Parameters window (figure 16-4, bottom). We'll add our parameters here. In the Filter option, select **Model dimensions.** We want to define the two dimensions that we want to change. To do this requires that the model dimensions are visible. For the easiest way to display the dimensions, right-click the *Annotations* folder in the SolidWorks Manager window to open an associated pop-up menu, and select **Show Feature Dimensions.**

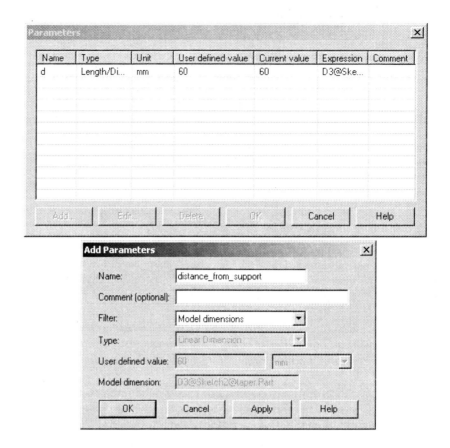

Figure 16-4: Parameters window and Add Parameters window

For this exercise, we call the parameter defining the distance from the center of the hole to the supported edge **distance_from_support**.

Having defined the parameter, we now define the Design Scenario. Right-click the *Design Scenario* folder to open a pop-up menu, and select **Edit/define**. Since we wish to change the hole position from 20 mm to 100 mm in 5-mm intervals, the number of scenarios is 17. Because more than one parameter can be defined in the Define design scenario area, the parameter (in our case, distance) must be selected (checked) as active as shown in figure 16-5.

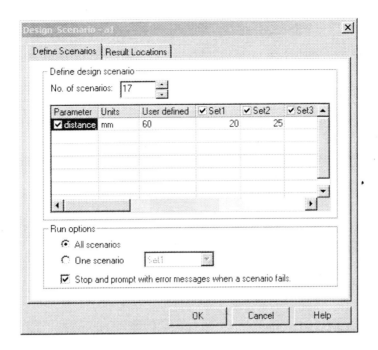

Figure 16-5: Define Scenarios tab in the Design Scenario window

Each one of 17 scenarios is distinguished by the particular value of the distance between the hole center and the supported edge, changing from 20 mm to 100 mm in 5-mm intervals. The name, a1, which appears at the top of this window is the study name.

To complete the definition of the Design Scenario, we need to inform COSMOSWorks what needs to be reported in those 17 steps. Select the Result Locations tab in the Design Scenario window, shown in figure 16-5, to access the screen shown in figure 16-6.

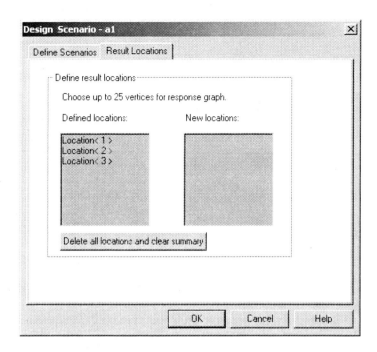

Figure 16-6: Result Locations tab in the Design Scenario – a1 window

*Only model vertexes can be selected as the desired location for the display of Design Scenario results. Defined locations can be renamed by right-clicking the item and selecting **Rename** from the associated pop-up menu.*

Now we can select the vertexes shown in figure 16-1. Since only vertexes are allowed as locations in the design scenario definition, it is now clear why split lines were added to the examined model geometry.

Run the design scenario by right-clicking the appropriate *Study* folder to open the associated pop-up menu and selecting **Run Design Scenario**. Once the run is complete, COSMOSWorks creates a *Design Scenario Results* folder, in addition to the other *Results* folders. Notice that results stored in all the other *Results* folders pertain to the geometry from the last step performed in the Design Scenario. The graphs in the *Design Scenario Results* folder are defined by right-clicking the *Design Scenario Results* folder to open the associated pop-up menu and selecting **Define Graph**. Figure 16-7 shows the Graph window where we decide which location(s) to display in the graph and what results they pertain to.

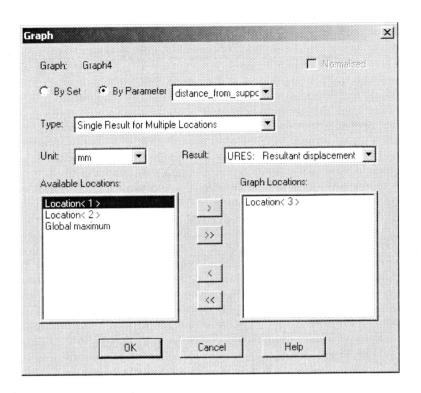

Figure 16-7: Graph window where locations for graphing are selected

*Desired location(s) to be graphed are selected in this window. Notice that **Global maximum** is also an available location.*

Figure 16-8 presents a sample of results with von Mises stress in location 1.

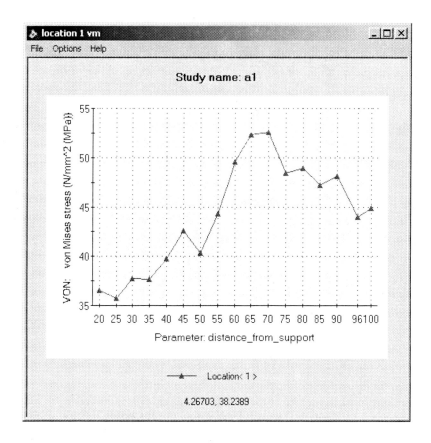

Figure 16-9: Von Mises stress as function of distance between the hole center and the supported face

Note that the maximum von Mises stress occurs between 65 mm and 70 mm. The parameter increments would have to be more refined to determine this maximum value more precisely.

You are encouraged to investigate the other options in the 2D Chart Control Properties window. They offer ample opportunities to manipulate graph display (figure 16-10).

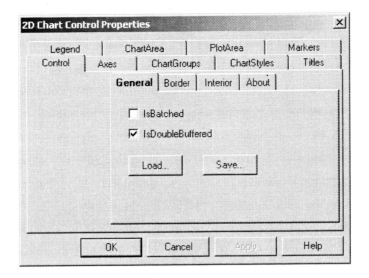

Figure 16-10: 2D Chart Control Properties window

The graph options allow customized graph display.

NOTES:

17: Miscellaneous topics

This chapter covers miscellaneous topics that add to your understanding of the capabilities of COSMOSWorks.

Selecting the automesher

You can select the Standard or Alternate automesher in the Preferences window under the Mesh tab (figure 2-15). The Standard automesher is the preferred choice. It uses the Voronoi-Delaunay meshing technique and is faster than the alternate automesher.

The Alternate automesher uses the Advancing Front meshing technique and should be used only when the Standard automesher fails even when you try various element sizes. The Alternate mesher ignores mesh control and automatic transition settings.

Mesh quality and mesh degeneration

The ideal shape of a tetrahedral element is a regular tetrahedron. The aspect ratio of a regular tetrahedron is assumed as 1. Analogically, an equilateral triangle is the ideal shape for a shell element. The further the shape departs from its ideal shape, the higher the aspect ratio becomes (figure 17-1). Too high an aspect ratio causes element degeneration and negatively affects the quality of the results provided by this element.

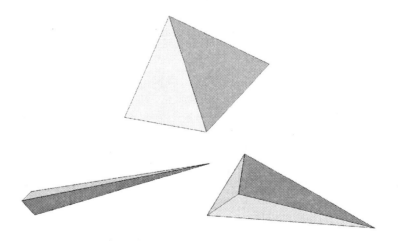

Figure 17-1: Tetrahedral element shapes

A tetrahedral element in the ideal shape (top) has as aspect ration of 1. "Spiky" and "flat" elements shown in this illustration (bottom) have excessively high aspect ratios.

The aspect ratio of a perfect tetrahedral element is used as the basis for calculating the aspect ratios of other elements. While the automesher tries to create elements with aspect ratios close to 1, the nature of geometry makes it sometimes impossible to avoid high aspect ratios (figure 17-2).

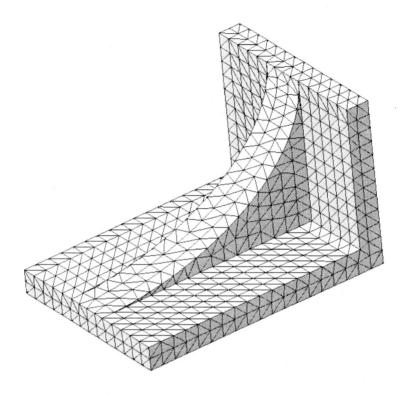

Figure 17-2: Mesh of an elliptical fillet

Meshing an elliptical fillet creates highly distorted elements near the tangent edges.

A failure diagnostic can be used to spot problem areas if meshing fails. To run a failure diagnostic, right-click the appropriate *Mesh* icon to open the associated pop-up window and select **Failure Diagnostic ….**However, meshing difficult geometries may sometimes result in degenerated elements without any warning. If mesh degeneration is only local, then we can simply not look at results (especially stress results) produced by those degenerated elements. If degeneration affects large portions of the mesh, then even global results can not be trusted.

Solvers and solvers options

Three solvers in combination with three solver options are available in COSMOSWorks (figure 17-3) even though not all solvers are available for all types of analyses (static, frequency, buckling, and thermal), and not all options are active with each solver. The default, and fastest, solver for most types of analyses is FFEPlus.

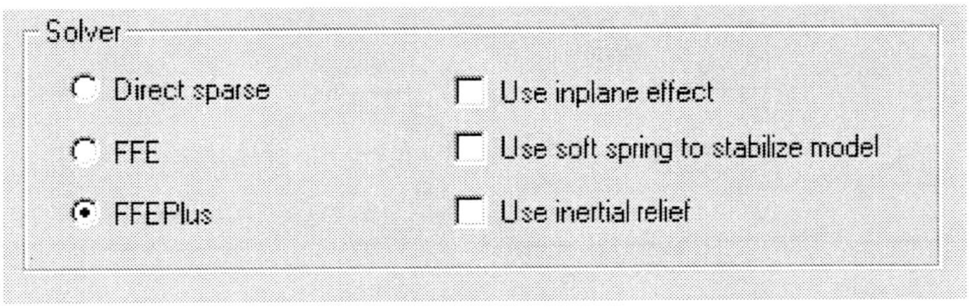

Figure 17-3: Solvers and solver options available in COSMOSWorks

The following summary recommends uses for each solver depending on the type of analysis:

Static analysis

Supported by all three solvers: Direct Sparse, FFE, and FFEPlus. The following is generally recommended:

- Direct Sparse for assembly problems with contact, especially with friction effects and when solving assemblies of parts with widely different material properties
- FFE or Direct Sparse for small and medium problems (problems with 100,000 to 300,000 degrees of freedom (DOF)
- FFEPlus for large and very large problems (problems with over 300,000 DOF)

Frequency analysis

Supported by two solvers: Direct Sparse and FFEPlus. There are two ways for solving frequency problems. You can use the mode extraction routine powered by the Direct Sparse solver or the FFEPlus solver.

Use:

- FFEPlus if rigid body modes are present
- Direct Sparse with **In plane effect** selected (checked) to consider pre-stress
- Direct Sparse to solve assemblies of parts with widely different material properties

Buckling problems

Use:

- FFEPlus to calculate the fundamental (first) mode of buckling
- Direct Sparse if more that one mode needs to be calculated

Thermal problems

Use:

- ❏ FFEPlus for large and very large problems
- ❏ Use Direct Sparse when solving assemblies of parts with widely different material properties

Solver options

Three solver options are available:

Option	Purpose
Use in plane effect	In a static analysis, use this option to account for changes in structural stiffness due to the effect of stress stiffening (when stresses are predominantly tensile) or stress softening (when stresses are predominantly compressive). In a frequency analysis, use this option to run a pre-stress frequency analysis (see chapter 18).
Use soft springs to stabilize the model	Use this option primarily to locate problems with restraints that result in rigid body motion. If the solver runs without this option selected and reports that the model is insufficiently constrained (an error message appears), the problem can be re-run with this option selected (checked). Insufficient restraints can then be detected by animating the displacement results. An alternative to using this option is to run a frequency analysis, identify the modes with zero frequency (these correspond to rigid body modes), and animate them to determine in which direction the model is insufficiently constrained.
Inertial relief	Use this option if a model is loaded with a balanced load, but no restraints. Because of numerical inaccuracies, the balanced load will report a non-zero resultant. This option can then be used to restore model equilibrium.

Contact/Gaps options

Several options are available in an assembly analysis when solving contact problems, i.e., when Contact/Gaps conditions are defined as **Node to Node**, **Surface**, or **Shrink Fit**. Those options are defined in the study properties, as shown in figure 17-4.

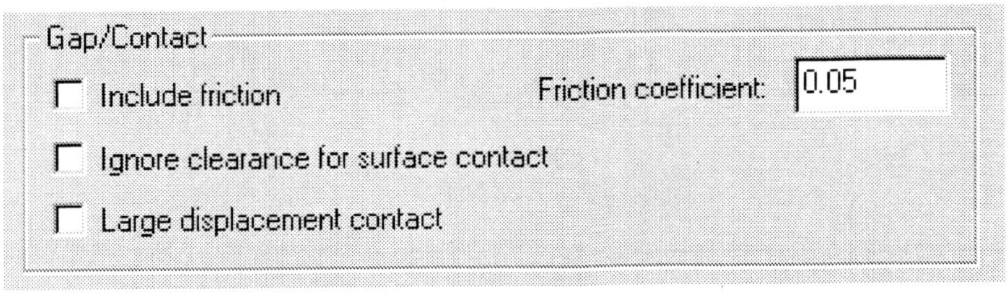

Figure 17-4: Contact/Gaps options available for contact stress analysis

The three options are described as follows:

Option	Purpose
Include friction	If selected (checked), friction between contacted surfaces is considered
Ignore clearance for surface contact	Use this option to ignore the initial clearance that may exist between surfaces in contact. The contacting surfaces start interacting immediately without first canceling out the gap.
Large displacement contact	Use this option if contacting surfaces need significant displacement before contact is made. "Significant" means that linear or angular displacements are significant in comparison with the size of the contacting surfaces. When in doubt, use this option, but note that it is quite computationally intensive.

Displaying mesh in a results plot

The default brightness of Ambient light, defined in SolidWorks Manager in the Lighting folder, is usually too dark to display mesh, especially a high-density mesh. A readable display of mesh (figure 17-5) requires increasing the brightness of Ambient light.

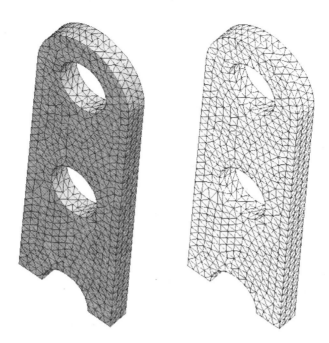

Figure 17-5: Mesh display in default Ambient light (left) and adjusted Ambient light (right)

A finite element mesh is displayed with ambient light brightness suitable for a CAD model (left) and with the brightness adjusted for displaying mesh (right).

Creating automatic reports

COSMOSWorks provides automated report creation. After a solution is complete and the desired Results plots have been created, right-click the *Report* folder (figure 17-6) to open the Report window.

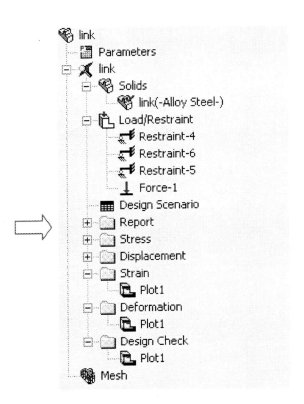

Figure 17-6: *Report* folder

*A **Report** folder may contain several reports. A report can be created only after analysis has been completed.*

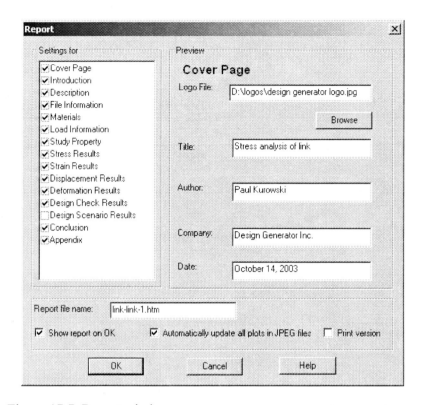

Figure 17-7: Report window

Right-click the Report *folder to open this window and specify desired report components.*

The report contains all plots from the *Results* folders that were selected (checked) in the Report window (figure 17-7) along with information on mesh, loads, restraints, etc.

Using e-drawings for result presentation

Each results plot can be saved in various graphic formats, as well as in SolidWorks eDrawing format. eDrawing format offers a very convenient way of communicating results analysis to users, who do not have COSMOSWorks.

Defining non-uniform loads

Although all the examples to this point have used uniformly distributed force or pressure, loads with non-uniform distribution can be defined easily.

We will illustrate this with an example of hydrostatic pressure acting on the walls of a 2-m deep tank, presented in the SolidWorks part file called NON UNIFORM LOAD. Note that the model uses meters for the unit of length. The pressure magnitude expressed in N/m^2 follows the equation p = 100,000 x, where x is the distance from the bottom of tank where coordinate system $cs1$ is located. The pressure definition requires selecting the coordinate system and the face where pressure is to be applied. The formula governing pressure distribution can be then typed in, as shown in figure 17-8.

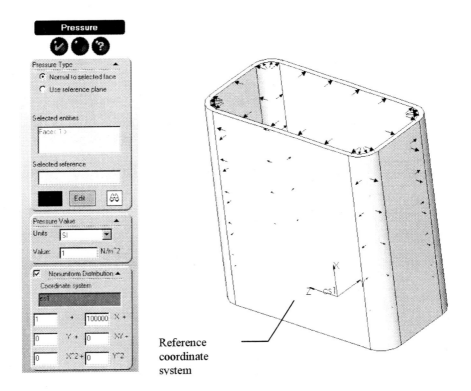

Figure 17-8: Water tank loaded with hydrostatic pressure requires linearly distributed pressure

Note that the vector length corresponds to pressure varying with x coordinate.

Defining bearing loads

The definition of a bearing load can be used to approximate contact pressure applied to a cylindrical face without modeling a contact problem. As seen in figure 17-9, the size of contact must be assumed (guessed) as indicated by the split face. The definition requires a coordinate system on which the z-axis is aligned with the axis of the cylindrical face. See the model in the SolidWorks file called BEARING LOAD for details.

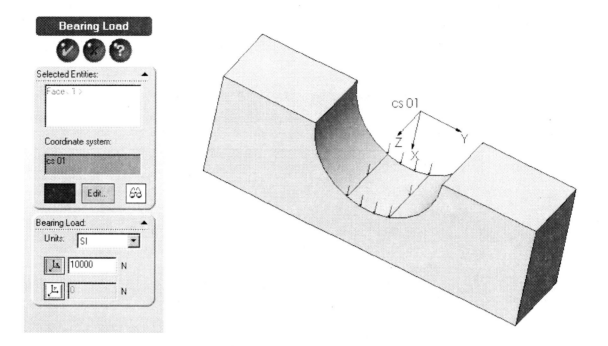

Figure 17-9: 10,000 N resultant force applied as pressure to a split face

Note that the pressure distribution follows the sine function

18: Selected advanced topics

The analysis capabilities of COSMOSWorks go far beyond those we have discussed so far. Readers are now sufficiently familiarized with this software to explore more advanced options and topics on their own. In this chapter, we provide a sample of less frequently used, but useful and interesting types of analyses. We selected the chapter title, Advanced Topic, for lack of a better name. While "advanced" certainly applies to the capabilities of analyses presented here, we do not mean to imply that those analyses are difficult to execute.

Frequency analysis with pre-load

A frequency, or modal, analysis of fast-rotating machinery requires that we account for stress stiffening. Stress stiffening is the increase in structural stiffness due to tensile loads. This commonly happens when components of rotating machinery are subjected to centrifugal load. We will illustrate this concept with the example of a turbine blade, which is located in the SolidWorks part file called TURBINE BLADE. In order to account for pre-load in a frequency analysis, click (check) the **Use in plane effect** option selected (checked) in the Properties window of the study that defines the frequency analysis (figure 18-1).

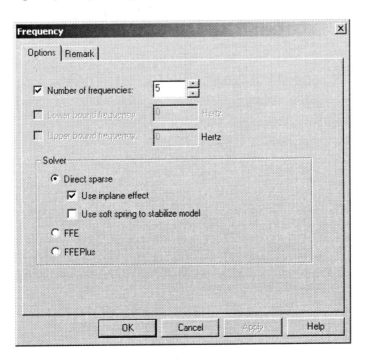

Figure 18-1: Frequency window with the **Use in-plane effects** selected

*When the **Use in-plane effects** option is selected, **Direct sparse** is the only available solver.*

Assign material properties of alloy steel and load the idealized turbine blade with centrifugal force that results from spinning at 15,000 RPM. To open the Centrifugal window (figure 18-2) and define the load, right-click the *Load/Restraints* folder to open a pop-up menu, and select **Centrifugal....** Having defined load proceed with support definition as shown in figure 18-3.

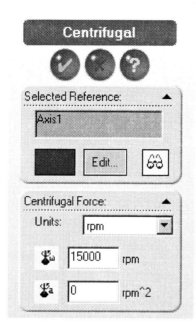

Figure 18-2: Centrifugal window with centrifugal force defined

Centrifugal force is applied as if the model were rotating about an axis, which is used as a reference in load definition. Notice that units are in revolutions per minute and that angular acceleration can also be defined if desired.

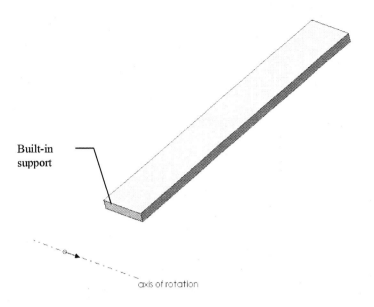

Figure 18-3: Idealized model of turbine blade showing built-in support

The idealized model of turbine blade is rigidly supported and loaded with centrifugal load.

Finally, mesh the model making sure to have two layers of 2nd order elements across the blade thickness, which is accomplished by selecting the proper global element size in the Mesh window.

To investigate the effect of pre-load, run the solution again with the **Use in plane effect** option deselected (unchecked).

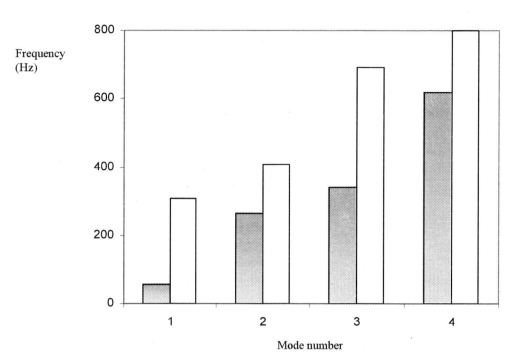

Figure 18-4: Comparison of frequencies of the first four modes with (dark bars) and without (light bars) in-plane effects.

The comparison of results graph, (figure 18-4) demonstrates the effect of stress stiffening on natural frequencies. Stress stiffness, which is developed by tensile stresses, adds to structural stiffness and results in higher natural frequencies. The opposite effect would be observed if, hypothetically, the turbine blade were subjected to compressive load.

For more practice, try conducting a frequency analysis of a beam under a compressive load. The higher the compressive load, the lower the first natural frequency. The magnitude of the compressive load, which causes the first natural frequency to drop to 0 (zero) is the buckling load. This is where frequency and buckling analyses meet!

Large deformations contact

If two surfaces, defined as in contact, experience large displacement before contacting each other, then the **Large deformation** option must be selected (checked) in the Properties window of the static study (17-4).

When exactly is displacement classified as "large"?

While there is no set rule, visible relative rotations or translations may have to be considered as large. To illustrate non-linear contact, let's examine the clip saved in the assembly file called LARGE DEF CONTACT (figure 18-5). Let's say we want to know the clip deformation under 100N force. Notice that even though the clip is just one part, we have had to split it into two parts and consider them as an assembly because the contacting faces must belong to different parts. Global Contact/Gaps conditions are left at the default setting: **Touching faces: Bonded**. Local Contact/gaps conditions between two surfaces likely to come in contact are defined as: **Surface**. Assign the material properties of alloy steel, and mesh the model to create two layers of elements across the thickness.

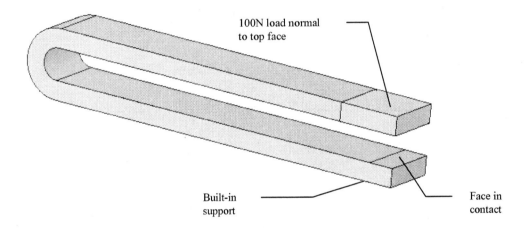

Figure 18-5: Clip modeled as an assembly because contacting faces must belong to different parts

Split lines in the SolidWorks model define faces in contact. The smaller size of the faces speeds up the solution.

Define two studies, one with the **Large deformation** option selected and the other without, and then compare the displacement results presented in figures 18-6 and 18-7.

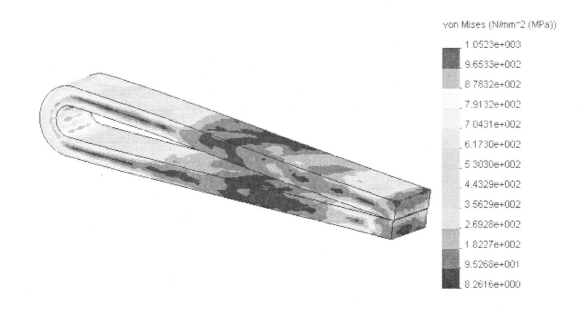

Figure 18-6: Erroneous stress results produced without the **Large deformation** option selected

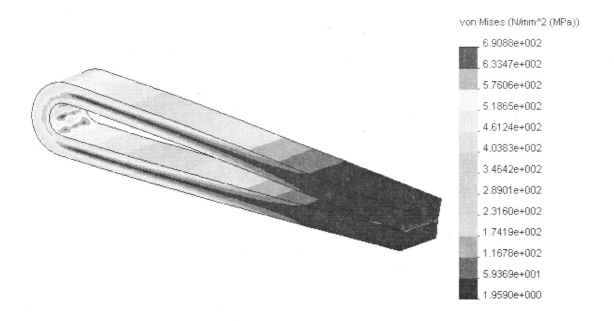

Figure 18-7: Correct stress results produced with the **Large deformation** option selected

The Stress results presented in figure 18-7 show the maximum von Mises stress at the bend. This is as far as we can carry the analysis of stress results. The element size is much too large in the contact area to produce meaningful contact stress results (figure 18-8).

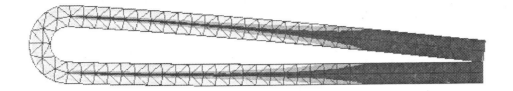

Figure 18-8: Large element size in the contact area prevents meaningful analysis of contact stresses

Shrink fit and inertial relief

Shrink fit is another type of Contact/Gaps condition that complements **Bonded, Free, Node to Node,** and **Surface** conditions. We will use it to analyze stresses developed as a result of press fit. The assembled parts are shown in figure 18-9; the exploded view along with information on component dimensions is shown in figure 18-10. Let's open the model geometry in the SolidWorks assembly file called SHRINK FIT.

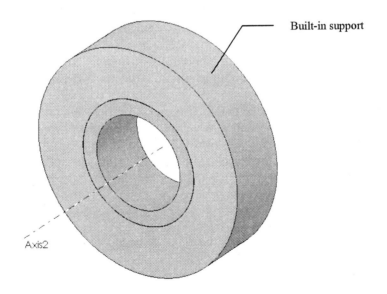

Figure 18-9: Small hollow ring pressed into a larger disk

Due to the press-fit, stresses will develop. Also shown is the axis, created in the SHRINK FIT assembly, which will be used as a reference for displaying radial stress results.

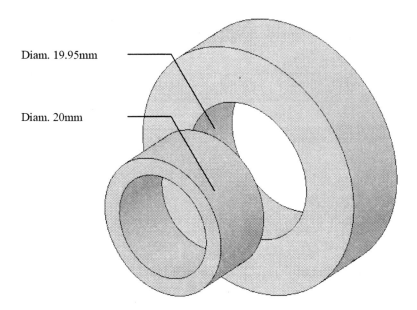

Figure 18-10: Interference of 0.05 mm between the two components

Components are shown here in exploded view

Assign material properties of AISI1020 steel and define Contact/Gaps conditions as **Shrink Fit**. Note that the mating conditions defined in SolidWorks assembly *do not* transfer to COSMOSWorks as Contact/Gaps conditions. Next, define Restraints as shown in figure 18-9. Note that this support is *not* sufficient to fully restrain the model. Remember that our contact problem ignores friction; therefore, the inside ring is free to spin and slide out. Note also that we are dealing with self-balanced loads. In this case we may use the option **Use inertial relief** found in Properties window (figure 17-3) of the study that we are going to run.

Results in the form of von Mises stresses and radial stresses are shown in figures 18-11 and 18-12. These results were produced using a mesh created with a global element size of 1.5 mm.

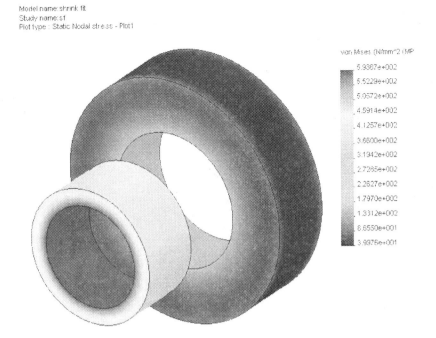

Figure 18-11: Von Mises stresses resulting from press fit

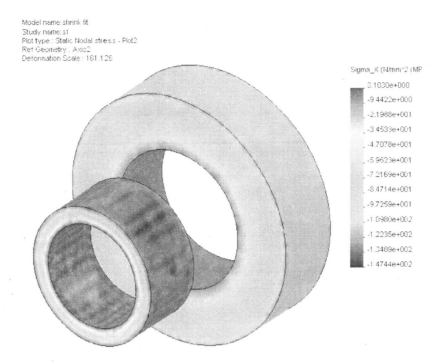

Figure 18-12: Radial stresses resulting from press fit

The axis shown in figure 18-9 is used as a reference to define this plot. Radial stresses can then be shown as stress in a global X direction.

19: Implementation of FEA into the design process

Positioning CAD and FEA activities

We have already stated that, from the point of view of design engineers, FEA is just another design tool. FEA should be implemented early in the design process and executed concurrently with design activities in order to help make more efficient design decisions. This concurrent CAD-FEA process is illustrated in figure 19-1.

Notice that design begins in CAD geometry and FEA begins in FEA-specific geometry. Every time FEA is used, the interface line is crossed twice: the first time when modifying CAD geometry to make it suitable for analysis with FEA and the second time when implementing results.

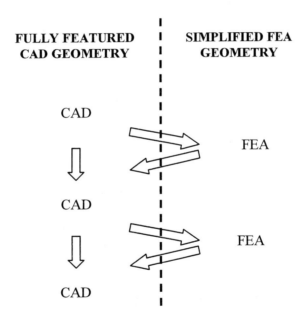

Figure 19-1: Concurrent CAD-FEA processes

CAD-FEA design process is developed in CAD-specific geometry while FEA analysis is conducted in FEA-specific geometry. Interfacing between the two geometries requires substantial effort and is prone to error.

The significant interfacing effort shown in figure 19-1 can be avoided if, recognizing the differences between CAD and FEA geometries, the new design is started and iterated in FEA-specific geometry. Only after running a sufficient number of iterations is the transfer to CAD geometry made by adding all manufacturing specific features. This way, the interfacing effort is reduced to just one switch from FEA to CAD geometry (figure 19-2).

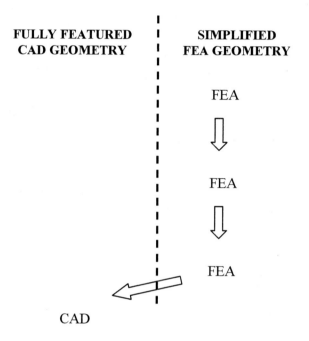

Figure19-2: FEA-driven design process

CAD-FEA interfacing efforts can be significantly reduced if the differences between CAD geometry and FEA geometry are recognized and the design process starts with FEA-specific geometry.

Major steps in an FEA project

Let's discuss the steps in an FEA project from a managerial point of view. While the previously described steps of creating a mathematical model, creating an FEA model, solution and analysis of results still apply, these must be appended by justification for analysis before any activity starts. After the analysis is complete, the results must be implemented. Those steps in an FEA project that require the involvement of management are marked with an asterisk (*).

Do I really need FEA? *

This is the most fundamental question to address before any analysis starts. FEA is expensive to conduct and consumes significant company resources to produce results. Therefore, each application should be well justified.

Providing answers to the following questions may help to decide if FEA is worthwhile:

- Can I use previous test results or previous FEA results?
- Is this a standard design so no analysis is necessary?
- Are loads, supports, and material properties known well enough to make FEA worthwhile?
- Would a simplified analytical model do?
- Does my customer demand FEA?
- Do I have enough time to implement the results of the FEA?

Should the analysis be done in house or should it be contracted out? *

Conducting analysis in-house *vs.* using an outside consultant has advantages and disadvantages. Consultants usually produce results faster while analysis performed in house is conducive to establishing company expertise leading to long-term savings.

The following list of questions may help in answering this question:

- How fast do I need to produce results?
- Do I have enough time and resources in-house to do complete a FEA before design decisions must be made?
- Is in-house expertise available?
- Is my software what my customer wants me to use?

Establish the scope of the analysis*

Having decided on the need to conduct FEA, we need to decide what type of analysis is required. The following list of questions may help in defining the scope of analysis.

Is this project:

- A standard analysis of a new product from an established product line?
- The last check of a production-ready new design before final testing?
- A quick check of design in-progress to assist the designer?
- An aid to an R&D project (particular detail of a design, gauge, fixture etc.)?
- A conceptual analysis to support a design at an early stage of development (e.g., R&D project)?
- A simplified analysis (e.g., only a part of the structure) to help making a design decision?

Other questions for consideration are:

- Is it possible to perform a comparative analysis?
- What is the estimated number of model iterations, load cases, etc.
- How should I analyze results? (applicable evaluation criteria, safety factors, and their values)
- How will I know whether the results can be trusted?

Establish a cost-effective modeling approach and define the mathematical model accordingly

Having established the scope of analysis, the FE model must now be prepared. The best analysis is, of course, the simplest one that provides the required results with acceptable accuracy. Therefore, the modeling approach should be as simple as possible to minimize project cost and duration, yet the approach should account for the essential characteristics of the analyzed object.

Here we need to decide on acceptable simplifications and idealizations to geometry. This decision may involve simplification of CAD geometry by defeaturing, or idealization by using shell representations. The goal is to produce a meshable geometry properly representing the analyzed problem.

Further, the definition of loads and supports must be formulated in accordance with the type of analysis to be performed. For example, a dynamic analysis requires loads and/or supports defined as a function of time.

Finally, material properties must also be defined in accordance with the type of analysis and degree of assumed simplifications.

Create a Finite Element model and solve it

The Finite Element model is created by discretization, or meshing, of a mathematical model. Although meshing implies that only geometry is discretized, discretization also affects loads and supports. Meshing and solution are largely an automated step, but it still requires input, which depending on the software used, may include:

- Element type(s) to be used
- Default element size and size tolerance
- Definition of mesh controls
- Automesher type to be used
- Solver type to be used

Review results

FEA results must be critically reviewed prior to using them for making design decisions. This critical review includes:

- ❏ Verification of assumptions and assessment of results (an iterative step that may require several loops to debug the model and to establish confidence in results)
- ❏ Study of the overall mode of deformations and animation of deflection to verify supports definition
- ❏ Check for Rigid Body Motions
- ❏ Check for overall stress levels (order of magnitude) using analytical methods in order to verify applied loads
- ❏ Check for reactions and construction of free body diagrams
- ❏ Review of discretization errors
- ❏ Analysis of stress concentrations and ability of the mesh to model them properly
- ❏ Review of results in difficult-to-model locations, such as thin walls, high stress gradients, etc.
- ❏ Investigation of the impact of element distortions on the data of interest

Analyze results*

The exact execution of this step depends, of course, on the objective of analysis. In all cases, however, results should be presented in a way suitable for using them to improve design.

- ❏ Present deformation results
- ❏ Present modal frequencies and associated modes of vibration
- ❏ Present stress results and corresponding factors of safety
- ❏ Consider modifications to the analyzed structure to eliminate excessive stresses and to improve material utilization and manufacturability
- ❏ Discuss results, and repeat iterations until an acceptable solution is found

Produce report*

- ❏ Produce report summarizing the activities performed, including assumptions and conclusions
- ❏ Append the completed report with a backup of relevant electronic files

Progress checkpoints in an FEA project

FEA project management requires the involvement of the manager involvement during project execution. The correctness of FEA results can not be established only by reviewing the analysis of results. A list of progress checkpoints may help a manager stay in the loop on a project and improve communication with the person performing the analysis. Several checkpoints are suggested in figure 19-3.

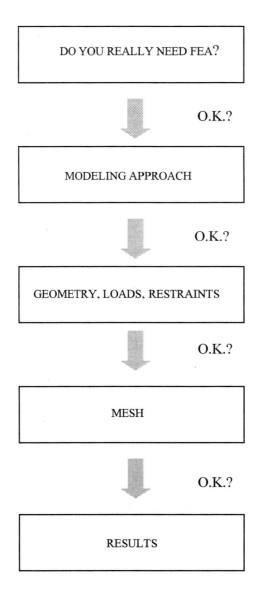

Figure 19-3: Checkpoints in an FEA project

The project is allowed to proceed only after the manager/supervisor has approved each step.

Structure of an FEA report

Even though each FEA project is unique, the structure of an FEA report follows similar patterns. Here are the major sections of a typical FEA report and their contents:

Section	Content
Executive Summary	Objective of the project, part number, project number, essential assumptions, results and conclusions, software used; including software release, information on where project backup is stored, etc.
Introduction	Description of the problem: Why did it require FEA? What kind of FEA? (static, contact stress, vibration analysis, etc.). What were the data of interest?
Geometry; loads and restraints	Description of model geometry and how it was created: from CAD geometry in integrated software, from CAD through an interface, built-in FEA software) Description and justification of any defeaturing and/or idealization Justification of the modeling approach (solids, shells) Description of loads and supports; include load diagrams (e.g., free body diagrams) Discussion of any simplifications and assumptions, etc.
Mesh	Description of the type of elements, global element size, any mesh control applied, number of elements, number of DOF, type of automesher used Justification of why this particular mesh is adequate to model the data of interest

Section	Content
Analysis of results	Presentation of displacement and stress results, including plots and animations
	Justification of the type of stress used to present results (e.g., max. principal, von Mises)
	Discussion of errors in the results and applicable safety factors considering analysis errors
	Discussion of the applicability of the safety factors in use
Conclusions	Recommendations to the requester re: structural integrity, necessary modifications, further study needed
	Recommendations for testing procedure (e.g., strain-gauge test, fatigue life test)
	Recommendations on future similar designs
Project documentation	Extensive documentation of design, design drawings, FEA model explanations, and computer back-ups
	Note that building in-house expertise requires very good documentation of the project beside the project report itself. Significant time should be allowed to prepare project documentation.
Follow-up	After completion of tests, report on test with test results appended
	Presentation of correlation between analysis results and test results
	Presentation of corrective action taken in case correlation is unsatisfactory (may involve revised model and/or tests)

20: Glossary

The following glossary provides definitions of terms used in this book.

Term	Definition
Adiabatic	An adiabatic wall is where there is no heat going in or out; it is perfectly isolated. In a thermal analysis, an adiabatic wall is one where no convection coefficients are defined.
Boundary Element Method	An alternative to the FEA method of solving field problems, where only the boundary of the solution domain needs to be discretized. Very efficient for analyzing compact 3D shapes, but difficult to use on more "spread out" shapes.
CAD	Computer Aided Design
Clean-up	Removing and/or repairing geometric features that would prevent the mesher from creating the mesh or would result in an incorrect mesh (with degenerated elements)
Constraints	Used in an optimization study, these are measures (e.g., stresses or displacements) that can not be exceeded during the process of optimization. A typical constraint would be the maximum allowed stress value.
Convergence criterion	Convergence criterion is a condition that must be satisfied in order for the convergence process to stop. In COSMOSWorks this applies to studies where the p-Adaptive solution has been selected. Technically, any calculated result can be used as a convergence criterion. The following convergence criteria can be used in COSMOSWorks: Total Strain Energy, RMS Resultant Displacement, and RMS von Mises stress.

Term	Definition
Convergence process	This is a process of systematic changes in the mesh in order to see how the data of interest change with the choice of the mesh and (hopefully) prove that the data of interest are not significantly dependent on the choice of discretization. A convergence process, or analysis, and be preformed as h-convergence or p-convergence.
	An h-convergence process is done by refining the mesh, i.e., by reducing the element size in the mesh and comparing the results before and after mesh refinement. Reduction can be done globally, by refining mesh everywhere in the model, or locally, by using mesh controls to refine the mesh only where stress concentrations are expected. An h-convergence analysis takes its name from the element characteristic, dimension h, which changes from one iteration to the next. An h-convergence analysis is performed by a user, who runs the solution, refines the mesh, compares results, etc.
	A p-convergence analysis, performed in programs supporting p-elements, does not affect element size, which stays the same throughout the entire convergence analysis process. Instead, element order is upgraded from one solution pass to the next. A p-convergence analysis is done automatically in an iterative solution until the user-specified convergence criterion is satisfied. A p-convergence analysis is done automatically. The only input required from the user is convergence criterion (or criteria) and desired accuracy.
	Sometimes, the desired accuracy can not be achieved even with the highest available p-element order. In this case, the user has to refine the p-element mesh manually in a fashion similar to traditional h-convergence, and then re-run the iterative p-convergence solution. This is called a p-h convergence analysis.
COSMOSDesignSTAR	This is FEA software that belongs to the family of COSMOS products and supports nonlinear analysis.

Term	Definition
Defeaturing	Defeaturing is the process of removing (or suppressing) geometric features from CAD geometry in order to simplify the finite element mesh.
Design scenario	In COSMOSWorks, this is an automated analysis of sensitivity of selected results to changes of a selected parameter defining the model, such as hole diameter, length of cantilever, etc.
Design variable	Used in an optimization study, this is a dimensional parameter that we wish to change within a specified range in order to achieve the specified optimization goal.
Discretization	This defines the process of splitting up a continuous mathematical model into discrete "pieces" called elements. A visible effect of discretization is the finite element mesh. However, loads and restraints are also discretized.
Discretization error	This type of error affects FEA results because FEA works on an assembly of discrete elements (mesh) rather than on a continuous structure. The finer the finite element mesh, the lower the discretization error, but the solution takes more time.
Element stress	This refers to stresses at nodes of a given element that are averaged between themselves (but not with stresses reported by other elements) and one value is assigned to the entire element. Element stresses produce discontinuous stress distribution in the model.
Element value	See Element stress.
FEA	Finite Element Analysis
Finite Difference Method	This is an alternative to the FEA method of solving a field problem, where the solution domain is discretized into a grid. The Finite Difference Method is generally less efficient for solving structural and thermal problems, but is often used in fluid dynamics problems.
Finite Element	Finite elements are the building blocks of a mesh, defined by position of their nodes and by functions approximating distribution of sought for quantities, such as displacements or temperatures.

Term	Definition
Finite Volumes Method	This is yet another alternative to the FEA method of solving field problem, similar to the Finite Difference Method
Frequency analysis	Also called modal analysis, a frequency analysis calculates the natural frequencies of a structure as the associated modes (shapes) of vibration. Modal analysis does *not* calculate displacements or stresses.
Gaussian points	These points are locations in the element where stresses are first calculated. Later, those stress results are extrapolated to nodes.
Gouraud display	This is a type of fringe plot display.
h element	h-elements are all elements, for which the order does not change during analysis. This means that a first order element remains a first order throughout solution. Convergence analysis of the model using h-elements is done by refining the mesh and comparing results (like deflection, stress, etc.) before and after refinement. The name, *h-element*, comes from the element characteristic dimension *h*, which is reduced in consecutive mesh refinements.
Idealization	This refers to making simplifying assumptions in the process of creating a mathematical model of an analyzed structure. Idealization may involve simplifying geometry, removing an entire half of the model and applying symmetry boundary conditions, replacing thin walls with mid-plane surfaces, etc.
Idealization error	This type of error results from the fact that analysis is conducted on an idealized model and not on a real-life object. Geometry, material properties, loads, and restraints all are idealized in models submitted to FEA.
Linear material	This is a type of material where stress is linearly proportional to strain.
Mesh diagnostic	This is a feature of COSMOSWorks that determines (in cases when meshing fails) which geometric entities prevented meshing
Meshing	This refers to the process of discretizing the model geometry. As a result of meshing, the originally continuous geometry is represented by an assembly of finite elements

Term	Definition
Modal analysis	See Frequency analysis.
Modeling error	This type of error results from idealizations that are introduced in the process of creating a mathematical model. See Idealization error.
Nodal stresses	These stresses are calculated at nodes by averaging stresses at a node as reported by all elements sharing that node. Nodal stresses are "smoothed out" and, by virtue of averaging produce continuous stress distributions in the model.
Node value	See Nodal stresses.
Numerical error	The accumulated rounding off of numbers causes this type of error by the numerical solver in the solution process. The value of numerical errors is usually very low.
Optimization criterion	See Optimization goal.
Optimization goal	Also called an optimization objective or an optimization criterion, the optimization goal is the objective of an optimization analysis. In an optimization study, for example, you could choose to minimize mass, volume, or frequency, or you could choose to maximize frequency or buckling (i.e., the load factor).
Optimization objective	See Optimization goal.
p-element	P-elements are elements that do not have pre-defined order. Solution of a p-element model requires several iterations while element order is upgraded until the difference in user-specified measures (e.g., global strain energy, RMS stress) becomes less than the requested accuracy. The name. *p-element*, comes from the p-order of polynomial functions (e.g., defining the displacement field in an element) which are gradually upgraded during the iterative solution along all edges.
p-Adaptive solution	This refers to an option available for static analysis with solid elements only. If p-Adaptive solution is selected (in the properties window of a static study), COSMOSWorks uses p-elements for iterative solution. A p-adaptive solution provides results with narrowly specified accuracy, but is very time-consuming and, therefore, impractical for large models.

Term	Definition
Pre-load	In this book, pre-load is a load that modifies the stiffness of a structure. Pre-load is important in a frequency analysis where it may significantly change natural frequencies.
Principal stress	This refers to stress that acts on the size of an imaginary stress cube in the absence of shear stresses. General 3D state of stress can be presented either by six stress components (normal stresses and shear stresses) expressed in an arbitrary coordinate system or by three principal stresses and three angles defining the cube orientation in relation to that coordinate system
Rigid body mode	This refers to a mode of vibration with zero frequency found in structures that are not fully restrained or not restrained at all. A structure with no supports has six rigid body modes. See Rigid body motion.
Rigid body motion	Rigid body motion is the ability to move without elastic deformation. In the case of fully supported structure, the only way it can move under load is to deform its shape. If a structure is not fully supported, or nor supported at all, it can move as a rigid body without any deformation.
RMS stress	Root Mean Square stress The name comes from the fact that it is the square root of the mean of the squares of the stress values in the model. RMS stress may be used as a convergence criterion if the p-adaptive solution method is used.
Sensitivity study	See Design scenario.
Shell element	Shell elements are intended for meshing surfaces of 3D or 2D models (Note that 2D models are not available in COSMOSWorks). The shell element that is used in COSMOSWorks is a triangular shell element. Triangular shell elements have three corner nodes. If this is a second order triangular element, it also has mid-side nodes, making then the total number of nodes equal to six. Each node of a shell element has 6 degrees of freedom. Quadrilateral shell elements are not available in COSMOSWorks.

Term	Definition
Small Deformations assumption	Analysis based on small deformations assumes that deformations caused by load are small enough as not to significantly change structure stiffness. Analysis based on this assumption of small deformations is also called linear geometry analysis or small displacement analysis. However, the magnitude of displacements itself is not the deciding factor in determining whether or not those deformations are indeed small or not. What matters, is whether or not those deformations significantly change the stiffness of the analyzed structure. For example, an initially flat membrane under pressure needs to deform only very slightly to significantly change its stiffness (by acquiring membrane stiffness), and therefore, requires a large deformation analysis, a function not available in COSMOSWorks, but available in COSMOSDesignSTAR. In COSMOSWorks, we assume that deformations are small.
SRAC	© 2003 Structural Research & Analysis Corp., SRAC are the creators of the family of COSMOS products. Phone: 800-469-7287; +1-310-207-2800 Email: info@srac.com
Steady state analysis	Steady state analysis in COSMOSWorks is an option in thermal analysis. It assumes that heat flow has stabilized and no longer changes with time
Structural stiffness	Structural stiffness is a function of shape, material properties, and restraints. Stiffness characterizes structural response to an applied load.
Symmetry boundary conditions	These refer to displacement conditions defined on a flat model boundary allowing for only for in-plane displacement and restricting any out-of-plane displacement components. Symmetry boundary conditions are very useful for reducing model size if model geometry, load, and supports are all symmetric. The model can then be cut in half and symmetry boundary conditions are applied to simulate the "missing half".

Term	Definition
Tetrahedral element	This is a type of element intended for meshing volumes of 3D models. It belongs to the family of solid elements beside hexahedral (brick) and wedge elements, which are not available in COSMOSWorks. A tetrahedral element has four triangular faces and four corner nodes. If this is used as a second order element (high quality in COSMOSWorks terminology) it also has mid-side nodes, making then the total number of nodes equal to 10. Each node of a tetrahedral element has 3 degrees of freedom
Thermal analysis	Thermal analysis finds temperature distribution and heat flow in a structure
Transient analysis	Transient analysis in COSMOSWorks is an option in a thermal analysis that plots how temperature distribution and heat flow change over time as a result of applied thermal loads and boundary conditions.
Tensile strength	This refers to the maximum stretching that can be allowed in a model before plastic deformation takes place.
Ultimate strength	This refers to the maximum stress that may occur in a structure. If the ultimate strength is exceeded, failure will take place (the part will break). Ultimate strength is usually much higher than tensile strength.
Von Mises stress	This is a stress measure that takes into consideration all six stress-components of a 3D state of stress. Von Mises stress, also called Huber stress, is a very convenient and popular way of presenting FEA results because it is a scalar non-negative value and because the magnitude of von Mises stress can be used to determine safety factors for materials exhibiting elasto-plastic properties, such as steel.
Yield strength	See Tensile strength.

21: FEA Resources

Many sources of FEA expertise are available to users. Sources include, but are not limited to:

- Engineering textbooks
- Software specific manuals
- Engineering journals
- Professional development courses
- FEA users' groups and e-mail exploders
- Government organizations

Engineering literature offers a large selection of FEA-related books, a few of which are listed here:

Adams V., Askenazi A. "Building Better Products with Finite Element Analysis", Onword Press, 1998.

Macneal R. "Finite Elements: Their Design and Performance", Marcel Dekker, Inc., 1994.

Spyrakos C. "Finite Element Modeling in Engineering Practice", West Virginia University Printing Services, 1994.

Szabo B., Babuska I. "Finite Element Analysis", John Wiley & Sons, Inc., 1991.

Zienkiewicz O., Taylor R. "The Finite Element Method", McGraw-Hill Book Company, 1989.